Grenades and Pyrotechnic Signals

FM 3-23.30

I0040707

HEADQUARTERS
DEPARTMENT OF THE ARMY

OCTOBER 2009

GRENADE,HAND,FRMG

COMP B LS-06C0

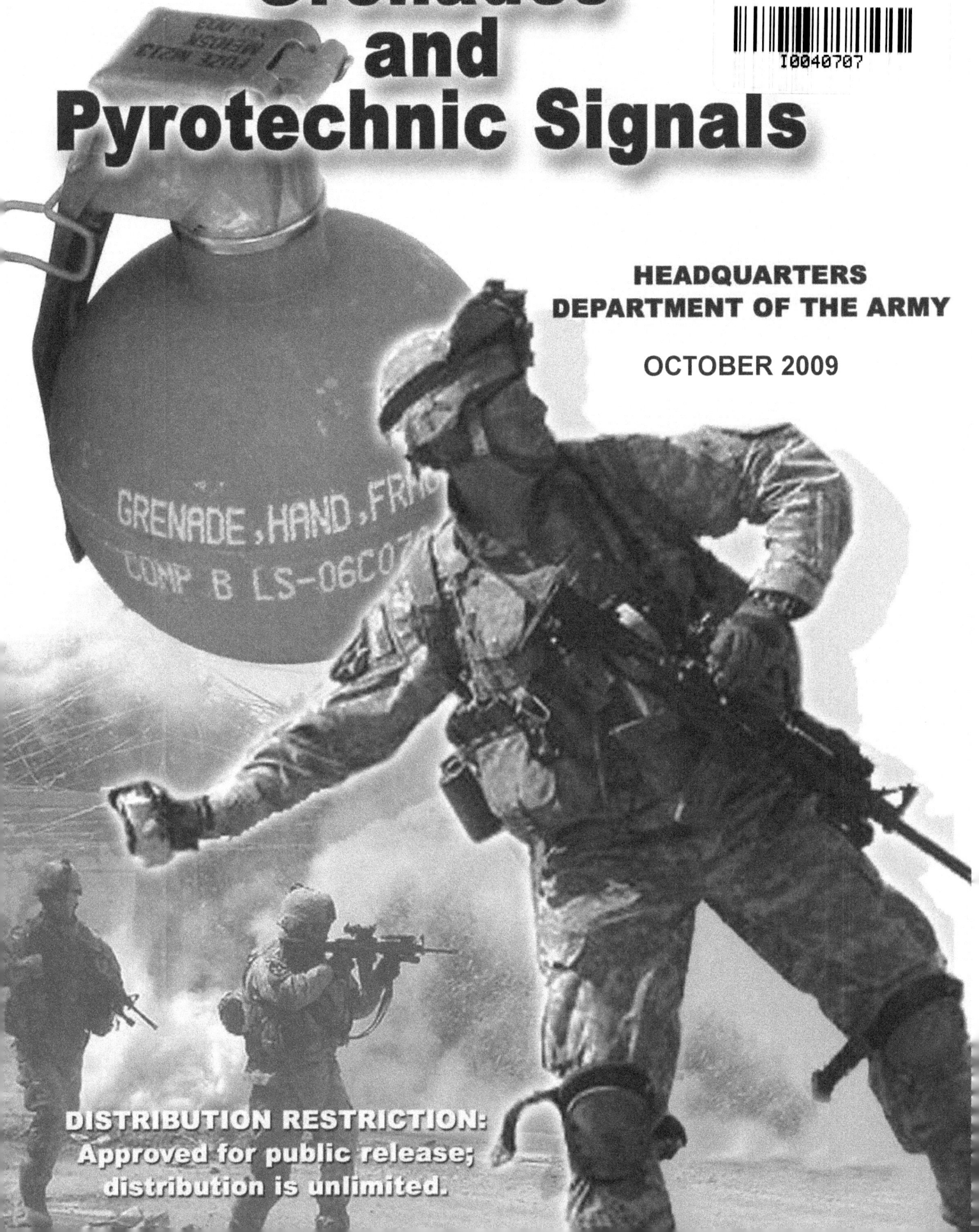

This publication is available at Army Knowledge Online (www.us.army.mil) and General Dennis J. Reimer Training and Doctrine Digital Library at (www.train.army.mil).

Field Manual
No. 3-23.30

Headquarters
Department of the Army
Washington, DC, 15 October 2009

Grenades and Pyrotechnic Signals

Contents

Page

Chapter 1 INTRODUCTION 1-1
Hand Grenades 1-1
Pyrotechnic Signals 1-2
Plans and Preparations for Training and Combat 1-3
Employment Rules 1-3

Chapter 2 TRAINING 2-1
Section I. Training Strategy 2-1
Objectives 2-1
Initial and Sustainment Training 2-1
Section II. Training Program 2-3
Mission-Essential Tasks 2-3
Training Assessment 2-3
Trainers 2-5
Trainer Certification Program 2-7
Section III. Training Preparation 2-8
Conduct a Training Risk Assessment 2-8
Conduct an Environmental Risk Assessment 2-10
Make Range Coordinations 2-10
Section IV. Training Conduct 2-27
Occupy, Inspect, and Set Up Range 2-27
Preparation for Training 2-28
Conduct the Training 2-28
Complete the Training Mission 2-43

Chapter 3 HAND GRENADES 3-1
Inspection 3-1
Storage 3-6
Use 3-9
Maintenance 3-19
Destruction Procedures 3-20

DISTRIBUTION RESTRICTION: Approved for public release; distribution is unlimited.

*This publication supersedes FM 3-23.30, 7 June 2005, and FM 3-23.30, Change 1, 27 November 2006.

Chapter 4   **PYROTECHNIC SIGNALS AND SIMULATORS** 4-1

Section I. Communication Signals 4-1

Handheld Signals 4-1

Maintenance 4-8

Destruction Procedures 4-8

Ground Smoke Signals 4-9

Inspection 4-9

Storage 4-9

Use 4-10

Section II. Trip Flares **4-11**

Storage 4-13

Use 4-13

Maintenance 4-15

Removal 4-15

Section III. Simulated Signals **4-15**

Early Warning Simulators 4-16

Ground-Burst Simulator 4-20

Hand Grenade Simulator 4-22

Section IV. Illumination Ground Signal Kits **4-23**

Inspection 4-23

Storage 4-24

Use 4-24

Maintenance 4-24

Destruction Procedures 4-24

Chapter 5   **EMPLOYMENT CONSIDERATIONS** **5-1**

Application 5-1

Rules of Engagement 5-2

Considerations 5-2

Offensive Operations 5-2

Defensive Operations 5-5

Retrograde Operations 5-5

Urban Operations 5-6

Air Operations 5-7

Use Under Adverse Conditions 5-7

Appendix A   **CHARACTERISTICS OF GRENADES** **A-1**

Fragmentation Hand Grenades A-1

Practice Hand Grenades A-2

Offensive Hand Grenades A-9

Nonlethal Hand Grenades A-10

Chemical Grenades A-11

Appendix B **CHARACTERISTICS OF PYROTECHNIC SIGNALS AND SIMULATORS** **B-1**
    Section I. Communication Signals.. **B-1**
        Handheld Signals...B-1
        Ground Smoke Signals..B-3
    Section II. Trip Flares **B-8**
    Section III. Simulated Signals.. **B-10**
        Early Warning Simulators B-10
        Ground-Burst Simulator B-13
        Hand Grenade Simulator B-14
    Section IV. Illumination Ground Signal Kits... **B-15**
        M185 and M186 Personnel Signal Kits...B-16
        M260 Red Personnel Distress Signal Kit..B-17

Figures

Figure 2-1. Initial entry training hand grenade training strategy. 2-2
Figure 2-2. Distance and accuracy layout. 2-11
Figure 2-3. Hand grenade mock-bay layout. 2-12
Figure 2-4. Throwing pit with knee wall. 2-13
Figure 2-5. Throwing pit with safety pits. 2-14
Figure 2-6. Suggested live-bay layout. 2-15
Figure 2-7. Hand grenade range requirements. 2-15
Figure 2-8. Hand grenade live-bay layout...2-16
Figure 2-9. Hand grenade live-bay layout, observation window. 2-16
Figure 2-10. Hand grenade qualification course layout. 2-17
Figure 2-11. Station 1, engage enemy from fighting position (standing). 2-18
Figure 2-12. Station 2, engage bunker (prone)..2-18
Figure 2-13. Station 3, engage enemy mortar position (kneeling). 2-18
Figure 2-14. Station 4, engage enemy from behind cover (prone). 2-19
Figure 2-15. Station 5, engage trench (standing). 2-20
Figure 2-16. Station 6, engage wheeled vehicle (kneeling)..2-20
Figure 2-17. Station 7, identify hand grenades and pyrotechnic signals. 2-21
Figure 2-18. Surface danger zone for live-bay. 2-26
Figure 2-19. Example of a completed DA Form 3517-R (Hand Grenade Qualification
    Scorecard) (front). 2-35
Figure 2-20. Example of a completed DA Form 3517-R (Hand Grenade Qualification
    Scorecard) (back)...2-36
Figure 2-21. Example of squad situational training exercise with hand grenades...............2-43
Figure 3-1. Hand grenade shipping container. 3-2
Figure 3-2. Hand grenade shipping canister...3-2
Figure 3-3. Hand grenade shipping canister with packing material. 3-3
Figure 3-4. Hand grenade with safety clip installed. 3-4
Figure 3-5. Safety clip installation..3-5

Figure 3-6. Hand grenade safety inspection points..3-6

Figure 3-7. Grenade carrying pouch (attached to enhanced tactical load-bearing vest)..........3-7

Figure 3-8. Grenade carrying pouch (attached to load-carrying equipment).........................3-7

Figure 3-9. Taping and storing hand grenades. 3-8

Figure 3-10. Taping hand grenades to Soldier gear...3-8

Figure 3-11. Right-hand grip. 3-9

Figure 3-12. Left-hand grip...3-10

Figure 3-13. Right-hand grip, removing the safety clip..3-10

Figure 3-14. Left-hand grip, removing the safety clip. 3-11

Figure 3-15. Pull ring grip, right-/left-hand thrower..3-11

Figure 3-16. Right-hand grip, pulling the safety pin..3-12

Figure 3-17. Left-hand grip, pulling the safety pin. 3-12

Figure 3-18. Standing position. 3-15

Figure 3-19. Prone-to-standing position. 3-16

Figure 3-20. Kneeling position...3-17

Figure 3-21. Prone-to-kneeling position. 3-18

Figure 3-22. Alternate prone position. 3-19

Figure 4-1. Handheld pyrotechnic signal. 4-1

Figure 4-2. Handheld signal shipping container..4-2

Figure 4-3. Handheld signal barrier bag..4-3

Figure 4-4. Handheld signal individual sealed steel container. 4-4

Figure 4-5. Handheld signal individual container. 4-4

Figure 4-6. Handheld signal removed from the individual container. 4-5

Figure 4-7. Safety inspection points—before storage. 4-5

Figure 4-8. Firing a handheld signal..4-7

Figure 4-8. Firing a handheld signal (continued)...4-8

Figure 4-9. Smoke grenades...4-9

Figure 4-10. M49A1 surface trip flare..4-11

Figure 4-11. M49A1 surface trip flare trigger spring position. 4-12

Figure 4-12. Early warning simulator shipping container and barrier bags. 4-16

Figure 4-13. Early warning simulator cardboard shipping box. 4-17

Figure 4-14. Early warning simulator removed from the cardboard shipping box.......................4-17

Figure 4-15. Kit components. 4-18

Figure 4-16. M115A2 ground-burst simulator...4-21

Figure 4-17. M116A1 hand grenade simulator..4-22

Figure 4-18. M260 illumination ground signal kit..4-23

Figure A-1. Confidence clip..A-1

Figure A-2. M67 fragmentation hand grenade. A-2

Figure A-3. M69 practice hand grenade...A-3

Figure A-4. M228 detonating fuze. A-4

Figure A-5. Gripping the M228 detonating fuze. A-5

Figure A-6. Placing the confidence clip over the threaded end of the M228 fuze. A-5

Figure A-7. Insert the M228 fuze into the M69 practice hand grenade body and
secure pull ring to the confidence clip (right hand)...................................A-6

Figure A-8. Securing the pull ring to the confidence clip (left hand). A-7

Figure A-9. M102 practice stun hand grenade. A-8

Figure A-10. M240 detonating fuze. A-9

Figure A-11. MK3A2 offensive grenade...A-10

Figure A-12. M84 stun grenade. A-11

Figure A-13. AN-M14 TH3 incendiary hand grenade. A-12

Figure A-14. ABC-M7A2 and ABC-M7A3 riot-control hand grenades. A-13

Figure B-1. Confidence clip. B-1

Figure B-2. Handheld pyrotechnic signal. B-2

Figure B-3. M18 colored smoke hand grenades..B-3

Figure B-4. M83 TA white smoke hand grenade. B-4

Figure B-5. AN-M8 HC white smoke hand grenade. B-5

Figure B-6. M106 white smoke hand grenade...B-7

Figure B-7. M49A1 surface trip flare...B-9

Figure B-8. M117 flash explosive booby trap simulator. B-10

Figure B-9. M118 illuminating explosive booby trap simulator. B-11

Figure B-10. M119 whistling booby trap simulator..B-12

Figure B-11. M115A2 ground-burst simulator...B-13

Figure B-12. M116A1 hand grenade simulator...B-15

Figure B-13. M185-red/M186-various color personnel signal kit. B-16

Figure B-14. M260 red personnel distress signal kit...B-17

Tables

Table 2-1. Officer in charge/range safety officer requirements..2-22

Table 2-2. Status of dropped grenade. 2-27

Table 2-3. Distance and accuracy course—task, condition, and standard..........................2-29

Table 2-4. Hand grenade qualification course stations...2-34

Table 2-5. Identify hand grenade and pyrotechnic signals—task, condition, and standard.......2-37

Table 2-6. Inspect and maintain hand grenades and pyrotechnic signals— task,
condition, and standard..2-37

Table 2-7. Employ a pyrotechnic signal—task, condition, and standard. 2-38

Table 2-8. Distance and accuracy course—task, condition, and standard..........................2-38

Table 2-9. Bunker complex course—task, condition, and standard. 2-39

Table 2-10. Trench complex course—task, condition, and standard...................................2-40

Table 2-11. Building complex course—task, condition, and standard.................................2-41

Table 2-12. Mock-bay training—task, condition, and standard...2-41

Table 2-13. Live-bay training—task, condition, and standard...2-42

Table 5-1. Types of hand grenades and their applications..5-1

Table 5-2. Types of pyrotechnic signals and their applications. 5-1

Table 5-3. Hand grenade employment during urban operations. 5-6

Table A-1. Components and characteristics of M67 fragmentation grenade. A-2

Table A-2. Components and characteristics of M69 practice grenade. A-3

Table A-3. Components and characteristics of M102 practice stun grenade. A-8

Table A-4. Components and characteristics of MK3A2 offensive grenade. A-10

Table A-5. Components and characteristics of M84 stun grenade. A-11

Table A-6. Components and characteristics of AN-M14 TH3 incendiary hand grenade. A-12

Table A-7. Components and characteristics of ABC-M7A2 and ABC-M7A3
      riot-control hand grenades. A-13

Table B-1. Handheld signal identification. B-2

Table B-2. Components and characteristics of M18 colored smoke hand grenade. B-4

Table B-3. Components and characteristics of M83 TA white smoke hand grenade. B-5

Table B-4. Components and characteristics of AN-M8 HC white smoke hand grenade. B-6

Table B-5. Components and characteristics of M106 white smoke hand grenade. B-7

Table B-6. Components and characteristics of M49A1 surface trip flare. B-9

Table B-7. Components and characteristics of M117 flash explosive booby trap
      simulator. B-11

Table B-8. Components and characteristics of M118 illuminating explosive booby
      trap simulator. B-12

Table B-9. Components and characteristics of M119 whistling booby trap simulator. B-13

Table B-10. Components and characteristics of M115A2 ground-burst simulator. B-14

Table B-11. Components and characteristics of M116A1 ground-burst simulator. B-15

Table B-12. Components and characteristics of M185-red/M186-various color
      personnel signal kit. B-17

Table B-13. Components and characteristics of M260 red personnel distress signal kit. B-18

Preface

The purpose of this manual is to orient Soldiers to the functions and descriptions of hand grenades and pyrotechnic signals. It also provides a guide for the proper handling and throwing of hand grenades and pyrotechnic signals, suggests methods and techniques for the tactical employment of hand grenades and pyrotechnic signals, and provides a guide for leaders conducting hand grenade and pyrotechnic signal training.

This manual is organized to lead the trainer through the material needed to conduct training during initial entry training (IET) and unit sustainment training. Preliminary subjects include discussion on the hand grenade and pyrotechnic signal's capabilities, mechanical training, and the fundamentals and principles of employing hand grenades and pyrotechnic signals. Live-fire applications are scheduled after the Soldier has demonstrated preliminary skills.

This manual was revised to include references to new materiel. This revision includes—

- Improvements that were made to hand grenades (addition of the confidence clip) and the M106 white smoke hand grenade.
- Warnings against taping grenades to Soldier's gear.
- Removal of the M8 smoke.
- Removal of the word "training" from the M83 white smoke, which will replace the M8 for tactical and training use.
- Changes to the hand grenade qualification scorecard.

This publication prescribes DA Form 3517-R (Hand Grenade Qualification Scorecard).

This publication applies to the Active Army, the Army National Guard (ARNG), National Guard of the United States (ARNGUS), and the US Army Reserve (USAR) unless otherwise stated.

Terms that have joint or Army definitions are identified in both the glossary and the text. Terms for which FM 3-23.30 is the proponent FM are indicated with an asterisk in the glossary.

Uniforms depicted in this manual were drawn without camouflage for clarity of the illustration. Unless this publication states otherwise, masculine nouns and pronouns refer to both men and women.

The proponent for this publication is the US Army Training and Doctrine Command (TRADOC). The preparing agency is the United States Army Infantry School (USAIS). You may send comments and recommendations by any means (US mail, e-mail, fax, or telephone) as long as you use DA Form 2028 (Recommended Changes to Publications and Blank Forms) or follow its format. Point of contact information is as follows:

E-mail:	benn.29IN.229-S3-DOC-LIT@conus.army.mil
Phone:	Commercial: 706-545-8623
	DSN: 835-8623
Fax:	Commercial: 706-545-8600
	DSN: 835-8600
US Mail:	Commandant, USAIS
	ATTN: ATSH-INB
	6650 Wilkin Drive, Bldg 74, Rm 102
	Fort Benning, GA 31905-5593

This page is intentionally left blank.

Chapter 1

Introduction

Hand grenades and pyrotechnic signals rapidly degrade the enemy's detection, observation, and engagement capabilities, enhancing the maneuver and firepower capabilities of ground forces conducting dismounted operations inside restrictive terrain. They also provide the commander a non-lethal capability that contributes to increased protection.

This manual covers general purpose (GP) grenades and pyrotechnic signals. Soldiers employing hand grenades and pyrotechnic signals not covered in this manual should adhere to the storage, handling, and employment considerations set forth in this manual and in all associated technical manuals (TMs).

HAND GRENADES

1-1. Hand grenades play an instrumental role in increasing combat effectiveness and survivability. They can be used in all types of terrain and employed in most combat situations to—
*   Eliminate the threat of enemy soldiers in the open and entrenched within fortified positions.
*   Mark positions.
*   Conceal operations.
*   Surprise the enemy.
*   Equalize the threat.
*   Destroy or disable enemy equipment, when other weapons or munitions are not available or are in short supply.

1-2. There are five types of hand grenades:
*   Fragmentation.
*   Chemical.
*   Offensive.
*   Nonlethal.
*   Practice and training.

1-3. Each grenade offers a unique capability that provides the Soldier with various options to successfully complete any given mission.

FRAGMENTATION GRENADES

1-4. Historically, the most important type of hand grenade is the fragmentation grenade. These grenades are used to produce casualties by the high-velocity projection of fragments.

CHEMICAL GRENADES

1-5. Chemical grenades are used for incendiary purposes, screening, signaling, training, or riot control.

OFFENSIVE GRENADES

1-6. Offensive hand grenades (e.g., concussion grenades) are much less lethal than fragmentation grenades on an enemy in the open, but they are very effective against an enemy within a confined space.

NONLETHAL GRENADES

1-7.   Nonlethal grenades are used for diversionary purposes or when lethal force is not desired. Nonlethal munitions are designed to incapacitate personnel while minimizing fatalities, permanent injury to personnel, and collateral damage to property and the environment.

PRACTICE AND TRAINING GRENADES

1-8.   Practice and training grenades are for training personnel in use, care, and handling of service grenades.

PYROTECHNIC SIGNALS

1-9.   Pyrotechnics range from flares to signals to simulators. Pyrotechnic signals supplement or replace normal communication means, mark locations, chart enemy courses, and provide illumination for search and rescue missions.

1-10.   There are four types of pyrotechnic signals:
* Communication signals.
* Trip flares.
* Simulated signals.
* Illumination ground signal kits.

COMMUNICATION SIGNALS

1-11.   There are two classifications of pyrotechnic communication signals: handheld signals and ground smoke signals. Both type of signals come in varied color patterns. Soldiers can use these patterns to coordinate troop movements and, in the case of an emergency, designate pick-up points.

TRIP FLARES

1-12.   Surface trip flares can be used to—
* Provide early warning of infiltration of enemy troops or signaling.
* Illuminate an immediate area.
* Ignite fires.
* Identify firing ports.
* Force the enemy to withdraw.
* Destroy small, sensitive pieces of equipment (in the same manner as an incendiary grenade).

SIMULATED SIGNALS

1-13.   Some pyrotechnic simulators can be used to provide early warning signals and to illuminate the immediate area; however, they are primarily designed to imitate the sounds and effects of combat detonations during field training exercises.

ILLUMINATION GROUND SIGNAL KITS

1-14.   The pen gun flare supports the small-unit leader in fire control, maneuver, and initiating operations such as ambushes. These signals are also a component of air crewmen's survival vest and are used for distress signaling or to identify ground locations for aircraft.

PLANS AND PREPARATIONS FOR TRAINING AND COMBAT

1-15.   The commander establishes basic and combat loads of hand grenades and pyrotechnic signals. The combat load is not a fixed quantity; it can be altered as the situation dictates. Units vary the combat load depending upon the commander's analysis of the mission, enemy, terrain and weather, troops and support available, time available, civil considerations (METT-TC).

1-16. The factors for determining the combat load are—
- Unit's mission.
- Weight.
- Weapons tradeoff.
- Distribution.

UNIT'S MISSION

1-17. The most important factor in determining the combat load is the unit's mission, which influences the type and quantity of hand grenades or pyrotechnic signals needed.

WEIGHT

1-18. A hand grenade, for example, weighs close to one pound. Consequently, each grenade the Soldier carries adds a pound to his total load.

WEAPONS TRADEOFF

1-19. Soldiers cannot carry everything commanders would like them to take into battle. Commanders must consider the value of various weapons and munitions to determine which contribute the most to mission accomplishment.

DISTRIBUTION

1-20. Different types of hand grenades or pyrotechnic signals are required on all missions. Leaders should distribute the hand grenades and pyrotechnic signals selected for a mission among several Soldiers, if not among all of them. Further, leaders should distribute to each Soldier the hand grenades or pyrotechnic signals that are required for his assigned tasks.

EMPLOYMENT RULES

1-21. Before employing hand grenades or pyrotechnic signals, or when in areas where they are in use—
- Know where all friendly forces are located.
- Know the sector of fire.
- Use the buddy or team system.
- Ensure the projected arc of the grenade and pyrotechnic signal is clear of obstacles.

This page is intentionally left blank.

Chapter 2

Training

This chapter outlines the hand grenade and pyrotechnic signal training program. This training program progresses from fundamentals to advanced training, culminating with the integration of hand grenades and pyrotechnics into situational and field training exercises.

SECTION I. TRAINING STRATEGY

The goal of the hand grenade and pyrotechnic signal training program is to produce Soldiers who are proficient in using hand grenades and pyrotechnic signals for any tactical situation. The primary focus of the training program is making Soldiers aware of the types of hand grenades and pyrotechnic signals available, and their purposes, capabilities, and target applications. Throughout this training, safe handling and employment practices must be incorporated in all instructions and task execution to reduce injuries.

OBJECTIVES

2-1.   The hand grenade training program progresses using the crawl—walk—run methodology. The program advances from fundamental to advanced training, culminating with the integration of hand grenades into situational and field training exercises.

2-2.   Once Soldiers achieve proficiency, a sustainment program is implemented to maintain the high proficiency level. The following progressive training outline can be used or modified as needed:

- Instruction on—
  - Safety inspection and maintenance of hand grenades and pyrotechnic signals.
  - Visual identification of hand grenades and pyrotechnic signals, and classification by their purposes and capabilities.
- Instruction and practical exercises on—
  - Fundamentals of hand grenade gripping procedures, throwing techniques, and throwing positions.
  - Fundamentals of pyrotechnic signal procedures, and employment techniques.
- Practical exercises emphasizing—
  - Distance and accuracy of hand grenades using targets of different types at various ranges.
  - Placement of smoke and incendiary grenades, and early warning devices; communication signal launching procedures; and use of ground-burst simulators.
  - Negotiation of training courses that integrate hand grenades and pyrotechnic signals into buddy team movement techniques and multiple target engagements at various ranges.

INITIAL AND SUSTAINMENT TRAINING

2-3.   The training strategy for hand grenades begins in initial entry training (IET) and continues in the unit.

INITIAL TRAINING

2-4.   In IET, Soldiers learn how to inspect and maintain hand grenades, prepare for throwing, and throw from three positions (standing, kneeling, and prone). Soldiers are given instruction on M67 and M69 hand grenades, and receive a demonstration of the AN-M14 TH3 incendiary hand grenade and M18 and M83 smoke grenades.

Then, Soldiers demonstrate skill in preparing and throwing practice grenades. Once proficiency is achieved, they progress to throwing live hand grenades.

2-5.   IET training culminates in the Soldier's proficiency assessment, which is conducted on the hand grenade qualification course (consisting of seven stations). During this training, Soldiers master the art of using cover and concealment to assault enemy soldiers in the open, in trenches, in fighting positions, in bunkers, and against enemy vehicles. At the last station, Soldiers must identify the hand grenades and pyrotechnics shown and demonstrated throughout the course.

2-6.   Figure 2-1 shows the IET training strategy.

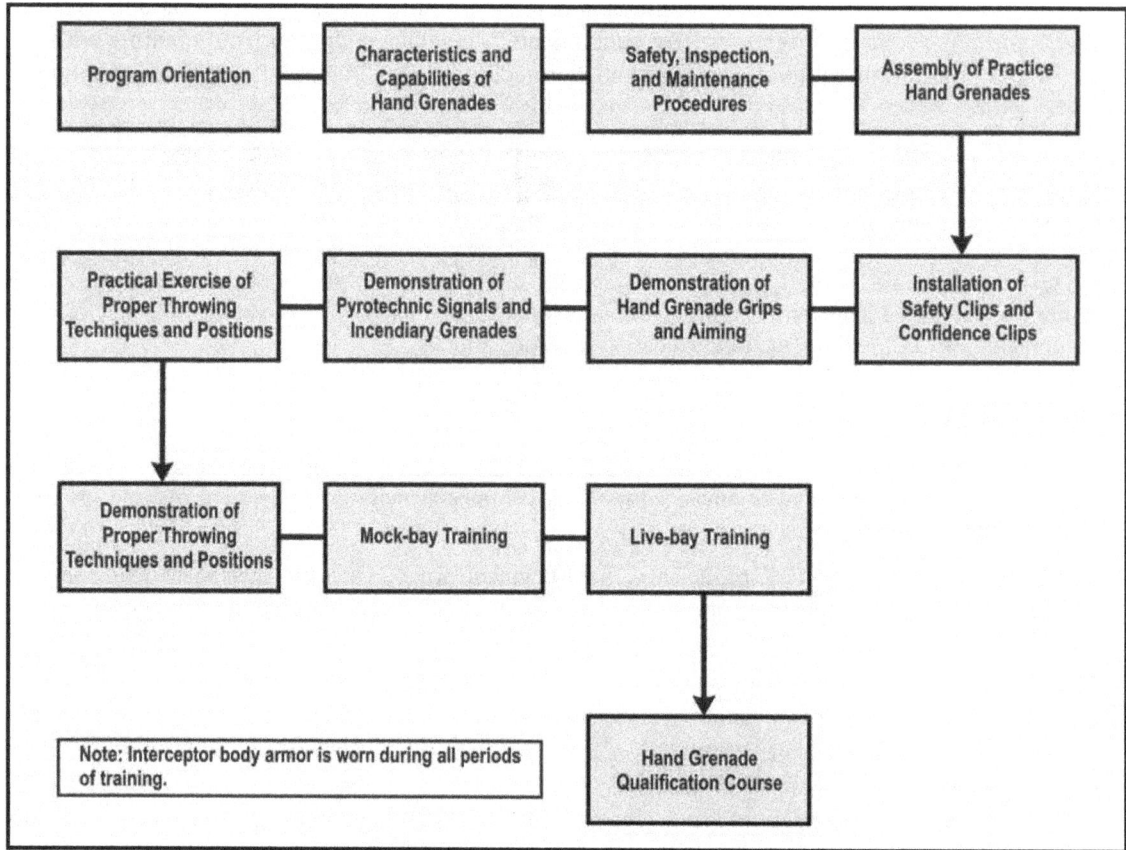

Figure 2-1. Initial entry training hand grenade training strategy.

SUSTAINMENT TRAINING

2-7.   Training continues in active Army, National Guard, and Army Reserve units using the same basic skills taught in IET, but at a higher level of skill. Units should set up a year-round program to sustain skills and have a plan for when they are at their home station and deployed.

NOTES: 1.   Hand grenade training is a high-risk form of individual training. Not only must units reinforce training Soldiers received during IET, they must add training on hand grenades used at the unit level (not covered in IET).

2.   Pyrotechnic signals are used throughout IET; however, IET Soldiers only view demonstrations provided by range instructors and training cadre. Unit leaders must ensure Soldiers are aware of the capabilities of pyrotechnic signals and are trained on their use before issuing them.

2-8. To sustain the basic hand grenade skills taught in IET, periodic preliminary instruction is conducted, followed by the hand grenade instructional and qualification course. Key elements include—

- Training the trainer.
- Refresher training of hand grenade skills using the M69 practice grenade.
- Sustainment training using live hand grenades.
- Remedial training.

NOTE: Not all Soldiers are proficient in throwing grenades. A Soldier who demonstrates high-risk tendencies during practice events must be identified and receive training reinforcement before throwing live hand grenades. Soldiers who continually demonstrate high-risk tendencies during reinforcement training should not be allowed to throw live hand grenades or use pyrotechnic signals.

2-9. Additional skills trained in the unit include—

- Hand grenades other than the M67.
- Pyrotechnic signal employment considerations.
- Mission-oriented protective posture (MOPP) with hand grenades and pyrotechnic signals.
- Employment of hand grenades (bursting and smoke) as a member of a fire team.

2-10. These skills are trained and integrated into collective training exercises, such as platoon and squad live-fire situational training exercises (STXs).

CAUTION

Fragmentation and concussion grenade training is considered high-risk training; these grenades should be thrown on designated hand grenade ranges only. Units must use the M69 with a M228 practice fuze and nonlethal hand grenades for reinforcement training and collective training scenarios.

SECTION II. TRAINING PROGRAM

The training listed in this chapter, except for the standard Army hand grenade qualification course, is offered as a model to assist units in meeting their training objectives. This training can be modified to support the unit mission-essential task list (METL), the terrain, and the commander's intent.

MISSION-ESSENTIAL TASKS

2-11. Hand grenade proficiency is critical to Soldiering and is required for any unit deployed to a wartime theater. All commanders should develop a METL and organize a training program that devotes adequate time to hand grenade and pyrotechnic signals. The unit's combat mission must be considered when establishing training priorities. This not only applies to the tasks selected for the unit's METL, but also to the conditions under which the tasks are to be performed, especially in an urban environment, where the effects of blast, burning, and/or vapor hazards, and wind cause adverse effects.

TRAINING ASSESSMENT

2-12. To conduct an effective hand grenade program, the unit commander must determine the current level of proficiency of all assigned personnel. Constant evaluation provides commanders understanding of where training emphasis is needed. All results are reviewed to determine any areas that need strengthening, along with any individuals that require special attention. Based on this evaluation, hand grenade and pyrotechnic signal training programs are developed and executed. Commanders continually assess the program and modify it as required. To develop a training plan and assess the training program, commanders should use the following tools:

- Direct observation of training.
- Spot checks.
- Review of past training.

2-13. Based on the commander's evaluation, goals, and missions, semiannual or annual training events are identified. Training programs must be continuous, and to sustain an effective program, resources are required. While the unit may only qualify its Soldiers annually, test results show that sustainment training is required at least semiannually to maintain Soldier skills.

DIRECT OBSERVATION OF TRAINING

2-14. Observing and accurately recording Soldier performance reveals the status of qualification results, and each Soldier's ability to identify the threat and successfully engage a target with the correct hand grenade. This also allows the commander to identify Soldiers who need special assistance to reach required standards and those who exceed these standards.

NOTE: Soldiers should also be certified on pyrotechnic signals.

SPOT CHECKS

2-15. Spot checks of individual training performance, such as interviews and evaluations of Soldiers, provide commanders with valuable information about Soldier proficiency and knowledge of the training tasks.

REVIEW OF PAST TRAINING

2-16. Commanders review past training to gain valuable information for developing a training plan. The assessment should include—

- The training program task.
- The frequency and results of training.

COMMANDER'S EVALUATION GUIDE

2-17. The commander's evaluation guide contains three sections:

- Commander's priorities and intent.
- Soldier assessment.
- Trainer assessment.

2-18. The following is an example of a commander's evaluation guide. Commanders can use this guide not only to assess their unit's training proficiency, but to assess the unit leaders and their ability to effectively implement a training program. They can also use it to develop noncommissioned officers (NCOs) into subject matter experts.

Commander's Priorities and Intent

2-19. When considering their priorities and intent, commanders answer the following questions:

- Have you clearly stated the priority of hand grenades and pyrotechnic signals in your unit? What is it? Do the staff and subordinates support this priority? Is it based on your METL and an understanding of FM 7-0?
- Have you clearly stated the intent of hand grenade and pyrotechnic signal training and qualification? Are leaders evaluating Soldier performance based on accurately recorded data and results?
- Have you clearly stated that hand grenade training and qualification is one of the commander's opportunities to assess several skills relating to team and squad readiness?

- What training and qualification course will be used to evaluate your unit's readiness?
    - How will it be conducted? Will the prescribed procedures be followed?
    - Who will collect the data?
- Have you clearly stated the purpose and intent of preliminary instruction?
    - What skills will preliminary instruction address?
    - Will preliminary instruction be performance-oriented? Are tasks integrated?

Soldier Assessment

2-20. During Soldier assessment, commanders answer the following questions:
- Do Soldiers know how to maintain, inspect, and stow their assigned hand grenades and pyrotechnic signals in accordance with the TM? Do they have a manual?
- Do Soldiers conduct serviceability checks of assigned hand grenades and pyrotechnic signals before training? Were maintenance deficiencies corrected?
- Do Soldiers demonstrate an understanding of the operation, functioning, and capabilities of hand grenades and pyrotechnic signals?
- Do Soldiers demonstrate their knowledge of the effects of wind when employing smoke grenades?
- During individual and collective training, do Soldiers demonstrate their ability to manage allocated hand grenades to engage all targets? Do they throw several hand grenades at one target?
- Do Soldiers demonstrate proficiency during night operations? When using night vision devices (NVDs)?
- Do Soldiers demonstrate individual proficiency during MOPP conditions? During collective exercises?
- Are hand grenade skills and pyrotechnic signals integrated into tactical exercises and unit live-fire exercises (LFXs)?
- Based on onsite observations and analysis of training performance, what skills or tasks show a readiness deficiency?
    - What skills need training emphasis? Individual emphasis? Leader emphasis?
    - What are the performance goals?

Trainer Assessment

2-21. During trainer assessment, commanders answer the following questions:
- Who has trained or will train the trainers?
    - What is the subject matter expertise of the cadre?
    - Are they actually training the critical skills?
    - Have they addressed the basic skills first?
    - What aids and devices are being used?
- What administrative constraints or training distracters can you overcome for the junior officer and NCO? Do the sergeants do the job they are charged with?
- At what level are the resources necessary to train hand grenades and pyrotechnic signals (time, training aids, ammunition, and ranges) controlled?

TRAINERS

2-22. Knowledgeable cadre/trainers are the key to hand grenade training performance. All commanders must be aware of maintaining expertise in hand grenade instruction/training.

CADRE/TRAINER

2-23. Cadre/trainer refers to a weapons instructor/trainer that has more experience and expertise than the Soldier who is provided the instruction. He trains Soldiers in the safe and effective use of hand grenades, and if required, pyrotechnic signals. The cadre/trainer maintains strict discipline during training, insists on compliance with range procedures and program objectives, and enforces safety regulations.

Selection

2-24. Institutional and unit cadre/trainers should be selected and assigned from the most highly qualified Soldiers. These Soldiers must demonstrate proficiency in all aspects of hand grenade and pyrotechnic signal employment, know the importance of training, and have a competent and professional attitude. The commander must ensure that selected unit cadre/trainers can effectively train other Soldiers. Local cadre/trainer training courses and weapons certification programs must be established to ensure that instructor/trainer skills are developed.

Duties

2-25. The cadre/trainer helps the Soldier master the fundamentals of hand grenade and pyrotechnic signal employment. He ensures that the Soldier consistently applies what he has learned. When training the beginner, the cadre/trainer confronts problems, such as fear, nervousness, forgetfulness, failure to understand, and a lack of coordination or determination, which may be compounded by arrogance or carelessness. With all types of Soldiers, the cadre/trainer must ensure that Soldiers are aware of their errors, understand the causes, and apply remedies. To perform these duties, cadre/trainers—

- Observe Soldier actions.
- Question the Soldier.
- Analyze Soldier actions.

Observing Soldier Actions

2-26. To pinpoint errors, the cadre/trainer observes the Soldier during drills, when preparing and throwing hand grenades, and when using pyrotechnic signals. If there is no indication of probable error, the Soldier's grip, preparation (removing safety devices or placement), launching/throwing position, release, and safe covering position are closely observed.

Questioning the Soldier

2-27. The Soldier is asked to state his throwing hand and to explain his throwing procedures.

Analyzing Soldier Actions

2-28. Analyzing Soldier actions is an important step in detecting and correcting errors. When analyzing Soldier actions, the cadre/trainer correlates observations of the Soldier to probable errors in performance, according to the type of hand grenade used and the target. A poor performance is usually caused by more than one observable error.

TRAINING THE TRAINER

2-29. Knowledgeable small-unit leaders are key to weapons training. This manual and other training publications provide the unit commander with the required information for developing a good train-the-trainer program.

2-30. The goal of a progressive train-the-trainer program is to achieve a high state of combat readiness. Through the active and aggressive leadership of the chain of command, a perpetual base of expertise is established and maintained.

2-31. The commander should identify unit personnel who have had assignments as weapons instructors. These individuals should be used to train other unit cadre by conducting preliminary instruction and LFXs for their Soldiers.

2-32. A suggested train-the-trainer program is outlined below:
- Conduct a diagnostic test to determine what training is needed.
- Conduct practice/live hand grenade range operations.
- Conduct hand grenade qualification course.

TRAINER CERTIFICATION PROGRAM

2-33. The certification program sustains the cadre/trainers' expertise and develops methods of training. The program standardizes procedures for certifying hand grenade and pyrotechnic signal trainers. Cadre/trainers' technical expertise must be continuously refreshed, updated, and closely managed.

TRAINING BASE

2-34. The training base can expect the same personnel changes as any other organization. Soldiers assigned as cadre/trainers will have varying degrees of experience and knowledge of training procedures and methods. Therefore, the trainer certification program must be an ongoing process that is tailored to address these variables. At a minimum, formal records should document program progression for each trainer. All cadre/trainers must complete the four phases of training using the progression steps, and the records should be updated on a quarterly basis.

CERTIFICATION PROGRAM OUTLINE

2-35. Before certification, all trainers must attend all phases of the train-the-trainer program in the following order:
- Phase I—Program Orientation.
- Phase II—Preliminary Hand Grenade and Pyrotechnic Signal Training.
- Phase III—Basic Hand Grenade Pyrotechnic Signal Training.
- Phase IV—Advanced Hand Grenade Pyrotechnic Signal Training.

2-36. Then, they conduct all phases to demonstrate their ability to train Soldiers and to diagnose and correct problems. Cadre/trainers who fail to attend or do not pass any phase of the diagnostic examination should be assigned to subsequent training.

Phase I—Program Orientation

2-37. During this phase, the cadre/trainer must accomplish the following tasks and be certified by the chain of command:
- Be briefed on the concept of the certification program.
- Be briefed on the unit weapons training strategy.
- Review the unit weapons training outlines.
- Review issued reference material.
- Visit training sites and firing ranges.

Phase II—Preliminary Hand Grenade and Pyrotechnic Signal Training

2-38. Phase II should be completed no more than two weeks following the conclusion of Phase I. During Phase II, the cadre/trainer demonstrates his ability to master hand grenade and pyrotechnic signal fundamentals, and his performance is reviewed by the chain of command. The results of this review are recorded and maintained on the trainer's progression sheet in accordance with the unit standing operating procedure (SOP). The cadre/trainer explains—
- Characteristics.
- Capabilities.
- Safety, inspection, and maintenance procedures.
- Assembly of practice hand grenades.

● Installation of safety clips and confidence clips (M228 practice fuze only).

NOTE: For hand grenades that come with a safety clip, the safety clip may detach during shipping and storage. See Chapter 3 for information about safety clip installation.

● Demonstration of pyrotechnic signals and incendiary grenades.
● Preparation (safety devices, right- and left-hand grips).
● Throwing positions and techniques.
● Mock-bay training.
● Live-bay training.
● Hand Grenade Qualification Course.

Phase III—Basic Hand Grenade and Pyrotechnic Signal Training

2-39. During this phase, the cadre/trainer demonstrates and reinforces what he has learned during Phase II. The cadre/trainer explains—

● Coordination requirements and range duties for conducting a hand grenade training course.
● The range layout and the conduct of training.

2-40. The commander determines when the cadre/trainer is ready to move to the next phase of certification. He will only do this when satisfied the cadre/trainer has successfully demonstrated expertise in setting up and conducting hand grenade and pyrotechnic signal training and qualification. At the completion of Phase III, the commander should schedule Phase IV certification and direct the cadre/trainer to begin coordinations. The results of this review should be recorded and maintained on the cadre/trainer's progression sheet.

NOTE: The commander should select a unit range control certified officer in charge (OIC) and range safety officer (RSO) to open and run the range so that he can view the cadre/trainer being certified.

Phase IV—Advanced Hand Grenade and Pyrotechnic Signal Training

2-41. The final phase of the train-the-trainer program tests the cadre/trainer. During this phase, the cadre/trainer sets up a hand grenade range and hand grenade qualification course, and conducts training for at least one person. If M69 hand grenades (with the M228 training fuze), M67 hand grenades, and pyrotechnic signals are available, the cadre/trainer conducts a firing exercise. If training ammunition and pyrotechnic signals are not available, the evaluation is based on the quality of training given.

SECTION III. TRAINING PREPARATION

Training preparation involves three steps:

(1) Conduct a training risk assessment.
(2) Conduct an environmental risk assessment.
(3) Make range coordinations.

CONDUCT A TRAINING RISK ASSESSMENT

2-42. The OIC or noncommissioned officer in charge (NCOIC) conducts a training risk assessment. It is vital to identify unnecessary risks by comparing potential benefit to potential loss. The composite risk management (CRM) process allows units to identify and control hazards, conserve combat power and resources, and complete the mission. This process is cyclic and continuous; it must be integrated into all phases of operations and training.

Application of the risk management process will not detract from this training goal, but will enhance execution of highly effective, realistic training.

FM 7-0, TRAINING THE FORCE

2-43. There are five steps to the CRM process:

(1) Identify hazards.

(2) Assess hazards to determine risk.

(3) Develop controls and make risk decisions.

(4) Implement controls.

(5) Supervise and evaluate.

NOTE: Risk decisions must be made at the appropriate level.

IDENTIFY HAZARDS

2-44. When identifying hazards, leaders should consider—

- The lethality of the hand grenades and pyrotechnic signals used.
- The area in which training is to be conducted.
- How the addition of new elements impacts known hazards.
- Any environmental impact.

ASSESS HAZARDS TO DETERMINE RISK

2-45. Once identified, hazards are assessed by considering the likelihood of its occurrence and the potential severity of injury without considering any control measures. When assessing hazards, leaders should consider the Soldiers' current state of training.

DEVELOP CONTROLS AND MAKE RISK DECISIONS

2-46. Leaders must apply two types of control measures to hand grenade risk assessments:

- Educational controls.
- Physical controls.

2-47. The unit commander's controls should be clear, concise, executable orders.

NOTE: Most vital to developing CRM controls is mature, educated leadership.

Educational Controls

2-48. Educational controls occur when adequate training takes place. They require the largest amount of planning and training time. Leaders implement educational controls using two sequential steps:

(1) Supervisors and instructors must be certified.

(2) Soldier training must be executed.

NOTE: Hand grenade and pyrotechnics training requires extensive direct supervision, but the amount of supervision required decreases as the Soldier's proficiency increases.

Physical Controls

2-49. Physical controls are the measures emplaced to reduce injuries. This includes not only protective equipment, but also certified personnel to supervise the training. Unrestrained physical controls are, in themselves, a hazard.

IMPLEMENT CONTROLS

2-50. When leaders implement the controls, they must match the controls to the Soldier's skill level. They must also enforce every control measure as a means of validating its adequacy.

SUPERVISE AND EVALUATE

2-51. This step allows leaders to eliminate unnecessary risk and ineffective controls by identifying unexpected hazards and determining if the implemented controls reduced the residual risk without interfering with the training.

CONDUCT AN ENVIRONMENTAL RISK ASSESSMENT

2-52. All leaders, trainers, and Soldiers must comply with environmental laws and regulations. The leader must identify the environmental risks associated with training individual and collective tasks, and implement environmental protection measures by integrating them into plans, orders, SOPs, training performance standards, and rehearsals.

2-53. Environmental risk management parallels safety risk management and is based on the same philosophy. Environmental risk management consists of identifying hazards before they happen and assessing hazards caused during training.

NOTE: See FM 7-0 for more information.

IDENTIFY HAZARDS

2-54. Leaders should identify the potential sources for environmental degradation during the analysis of METT-TC factors. An environmental hazard is a condition with the potential for polluting air, soil, or water or destroying cultural or historical artifacts.

ASSESS HAZARDS

2-55. Leaders should analyze the potential severity of environmental degradation by using the environmental risk assessment matrixes in FM 7-0. The risk effect value is defined as an indicator of the severity of environmental degradation. Leaders quantify the risk to the environment resulting from the operation as extremely high, medium, or low using the environmental assessment matrixes.

MAKE RANGE COORDINATIONS

2-56. Once the risks assessment is completed, viewed, and command approved, then the OIC or NCOIC should check out the range and coordinate for range use.

NOTE: The OIC or NCOIC should coordinate at least one day ahead of actual use to rehearse range setup and conduct.

RANGES

2-57. Ranges include distance and accuracy ranges, mock-bay throwing pits, and live-bay throwing pits.

Distance and Accuracy Ranges

2-58. A four-lane layout (Figure 2-2) is recommended. These lanes should enable Soldiers to engage—

- A fighting position at 30 meters.
- A trench target at 40 meters.
- A fortified mortar pit at 20 meters.
- Soldiers in the open at 20 meters.

NOTE: The four lanes may be combined if the terrain does not allow four stations.

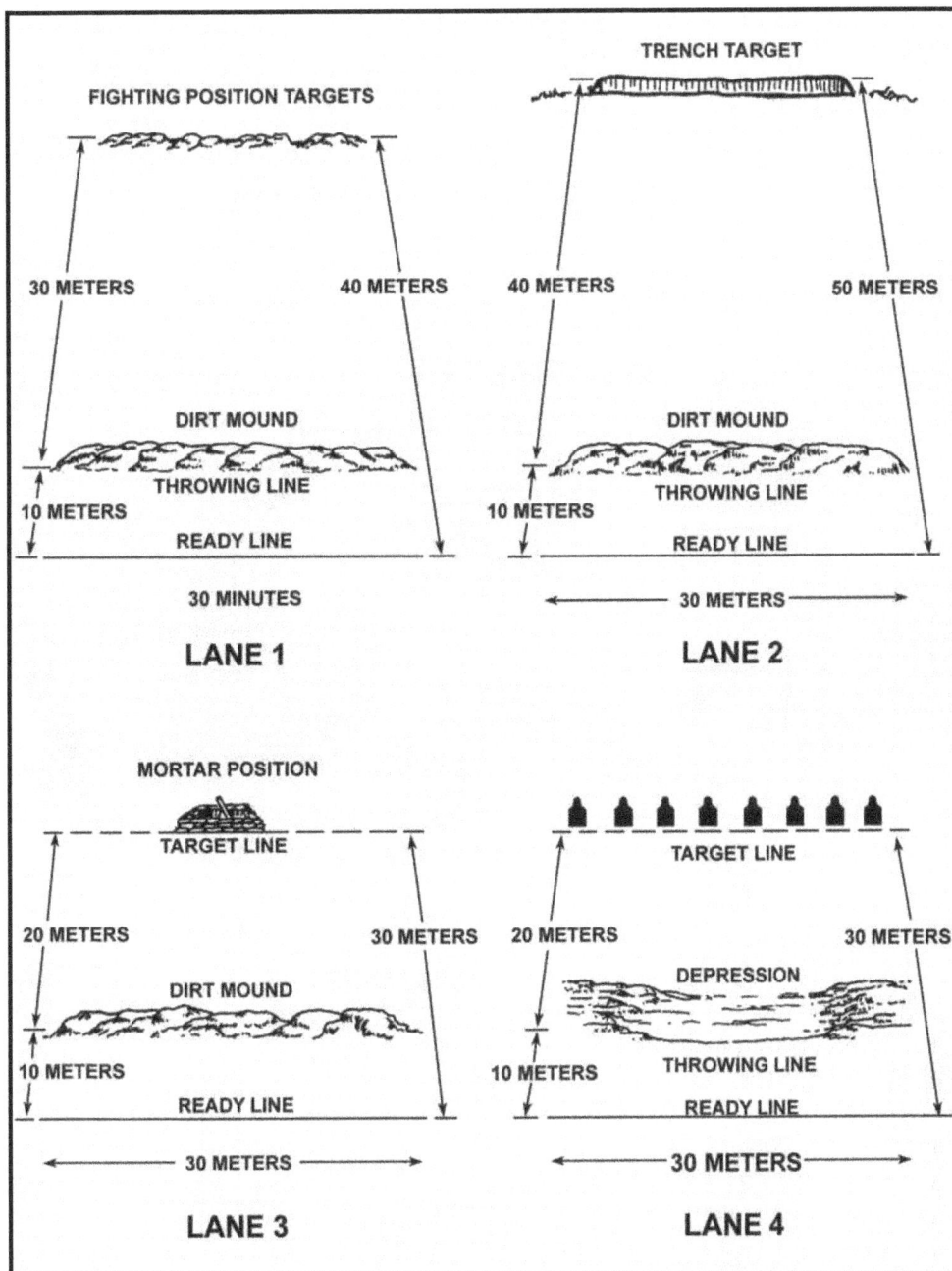

Figure 2-2. Distance and accuracy layout.

Mock-Bay Throwing Pits

2-59. The hand grenade mock-bay must replicate the dimensions and safety areas found at live-bay. At this station of hand grenade training, Soldiers are instructed on live-bay procedures. Each Soldier attending this training will have to identify his throwing hand and demonstrate the correct throwing procedures.

NOTE: Soldiers will throw from the standing position during mock-bay and live-bay training.

Throwing Pit and Knee Wall

2-60. A mock-bay training pit can be made from treated plywood (Figure 2-3). Protective reinforcement materials used in a live-bay throwing pit are not necessary.

Suspended Guide Wire

2-61. Place an approximately 6-meter (20-feet) high suspended guide wire approximately 28 meters (92 feet) in front of the mock-bay pit will assist Soldiers in achieving throwing distance.

E-Type Silhouettes

2-62. Placing E-type silhouettes at 40 meters from the mock-bay pit will provide targets for throwing accuracy and distance.

Figure 2-3. Hand grenade mock-bay layout.

Live-Bay Throwing Pits

2-63. Figures 2-4 through 2-9 depict a suggested hand grenade live-bay design. Live-bay throwing pits should incorporate the following elements:
- Observation pits or tower.
- Throwing pit.
- Knee wall.
- Sand/sawdust pit.
- Revetments and berms.
- Separation distance.
- Observation windows.

Observation Pits or Tower

2-64. Observation pits or towers should be a sufficient height to enable range personnel to observe and control all throwing pits. Laminated 35-millimeter (about 1 3/8 inches) windowpanes (constructed as described below) provide the necessary degree of safety:

- 10-mm glass (outside).
- 7-mm polycarbonate resin sheet.
- 6-mm glass.
- 6-mm polycarbonate resin sheet.
- 6-mm glass.

ThrowingPit

2-65. The throwing pit design is based on the average depth of the two-man foxhole (approximately 50 inches tall). Approximately armpit high. This height allows the Soldier to stand, see the target, and safely throw the hand grenade. The height of the wall, will also provide cover when in a kneel position.

Knee Wall

2-66. The throwing pit should have a rear wall (knee wall) no more than 0.6 meter (2 feet) high and 0.15 meter (6 inches) thick (Figure 2-4). The knee wall should extend the width of the throwing pit, connecting both ends of the enclosure. The top of the knee wall should slope inward to allow any grenade dropped on the wall to roll into the throwing pit. The knee wall should have drain pipes (no more than 2 inches in diameter) to allow throwing pit drainage. The floor of the pits should slope in the direction of the drainage pipes.

NOTE: DO NOT construct grenade sumps or ditches inside the throwing pits.

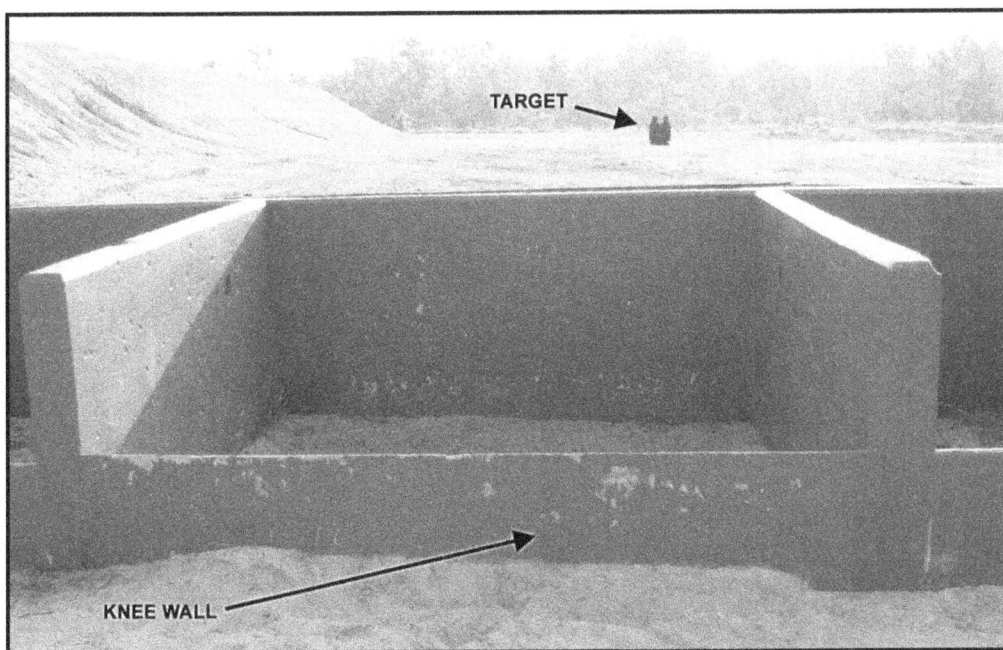

Figure 2-4. Throwing pit with knee wall.

2-67. Throwing pits that do not have knee walls must have safety pits attached to both sides (Figure 2-5).

Figure 2-5. Throwing pit with safety pits.

Sand/Sawdust Pit

2-68. A sand/sawdust pit is placed outside of the knee wall to cushion the fall of personnel diving over the wall in the event a grenade is dropped in the throwing pit.

Revetments and Berms

2-69. Where possible, the throwing pits should be separated using steel, concrete, or wooden revetments or earthen berms of a length and height to lessen the effect of high velocity, low-angle fragments (e.g., 50 meters long and 1.8 meters high). The thickness varies according to the type of construction used. This permits grenade throwing to continue from the adjacent pit when a dud grenade requires closure of a specific pit pending dud disposal.

Separation Distance

2-70. Live-bay throwing pits should have a separation distance of 25 meters between each lane. This places adjacent pits outside the effective casualty-producing radius of 15 meters for the M67 fragmentation grenade.

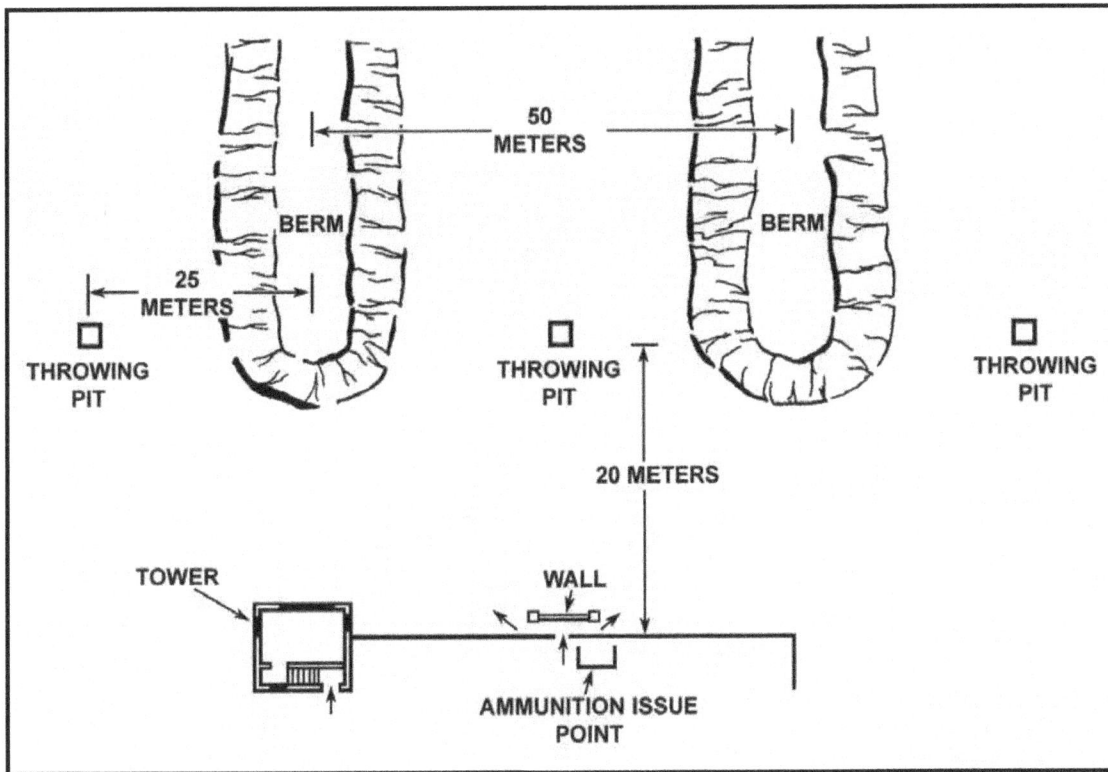

Figure 2-6. Suggested live-bay layout.

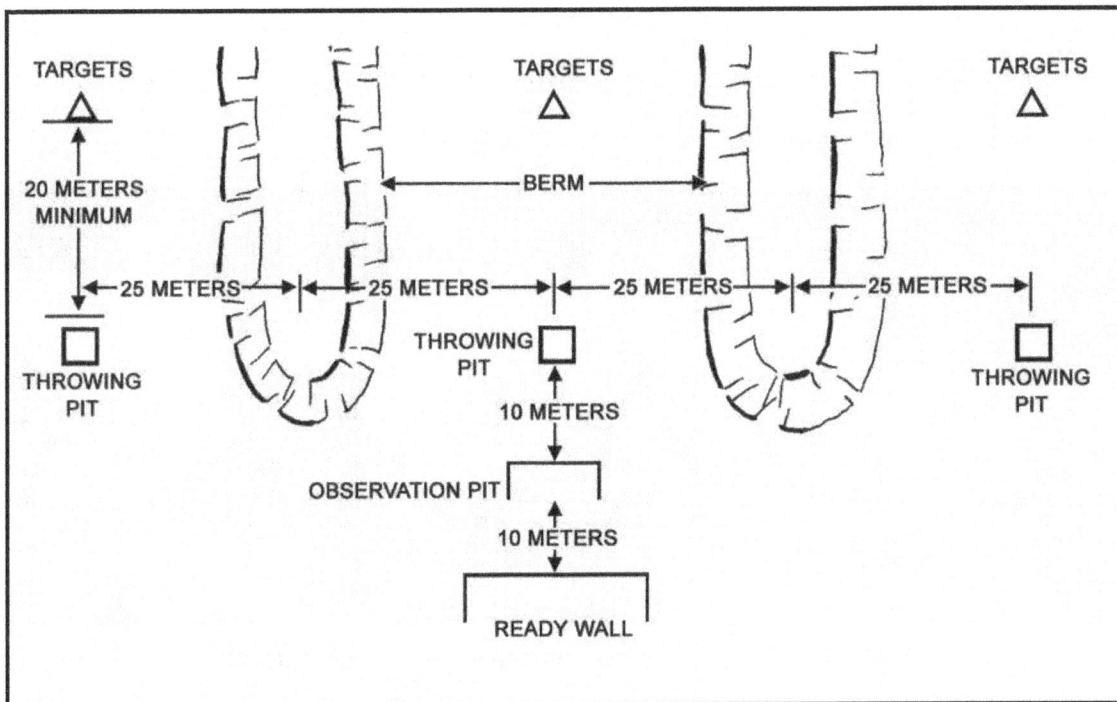

Figure 2-7. Hand grenade range requirements.

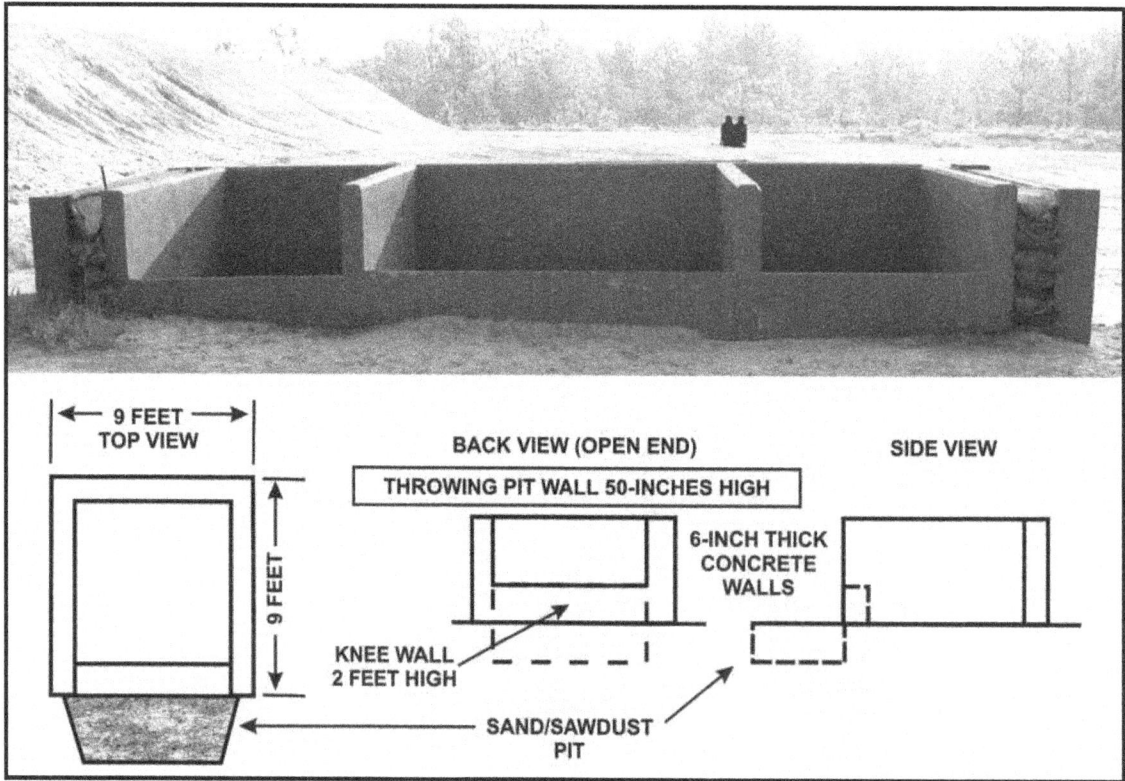

Figure 2-8. Hand grenade live-bay layout.

Observation Windows

2-71. If facilities permit, an observation window should be constructed to allow Soldiers to observe the live-bay throwing procedures before and after throwing the hand grenades (Figure 2-9). The observation window must be of the same construction used for observation pits and towers.

Figure 2-9. Hand grenade live-bay layout, observation window.

Hand Grenade Qualification Course

2-72. A seven-station layout (Figures 2-10 through 2-17) is recommended. These stations should enable Soldiers to perform the following tasks:

- Engage the enemy from a fighting position at 35 meters (standing).
- Engage bunker (prone).
- Engage enemy mortar position at 25 meters (kneeling).
- Engage enemy behind cover at 20 meters (prone).
- Engage trench at 25 meters (standing).
- Engage wheeled vehicle at 25 meters (kneeling).
- Identify hand grenades and pyrotechnic signals.

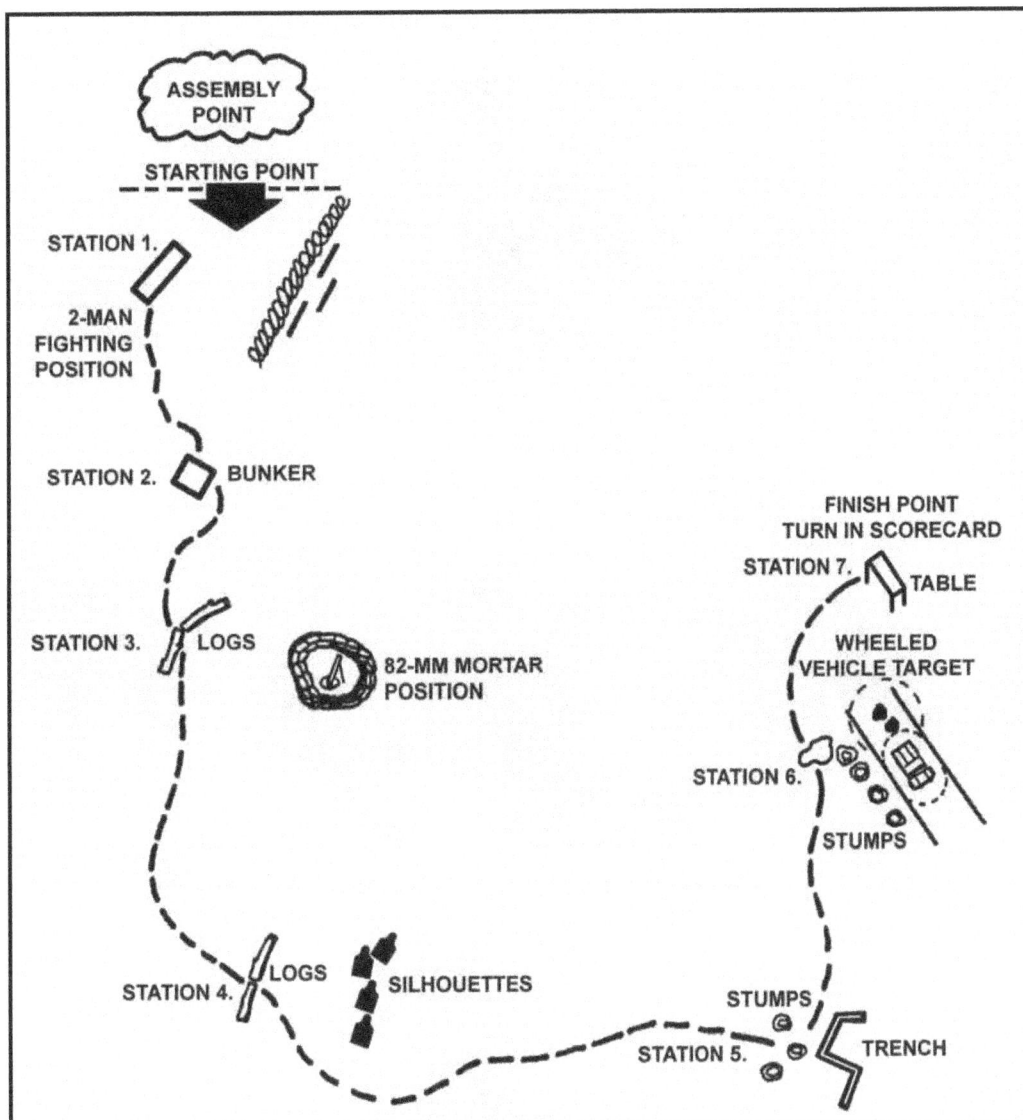

Figure 2-10. Hand grenade qualification course layout.

Figure 2-11. Station 1, engage enemy from fighting position (standing).

Figure 2-12. Station 2, engage bunker (prone).

Figure 2-13. Station 3, engage enemy mortar position (kneeling).

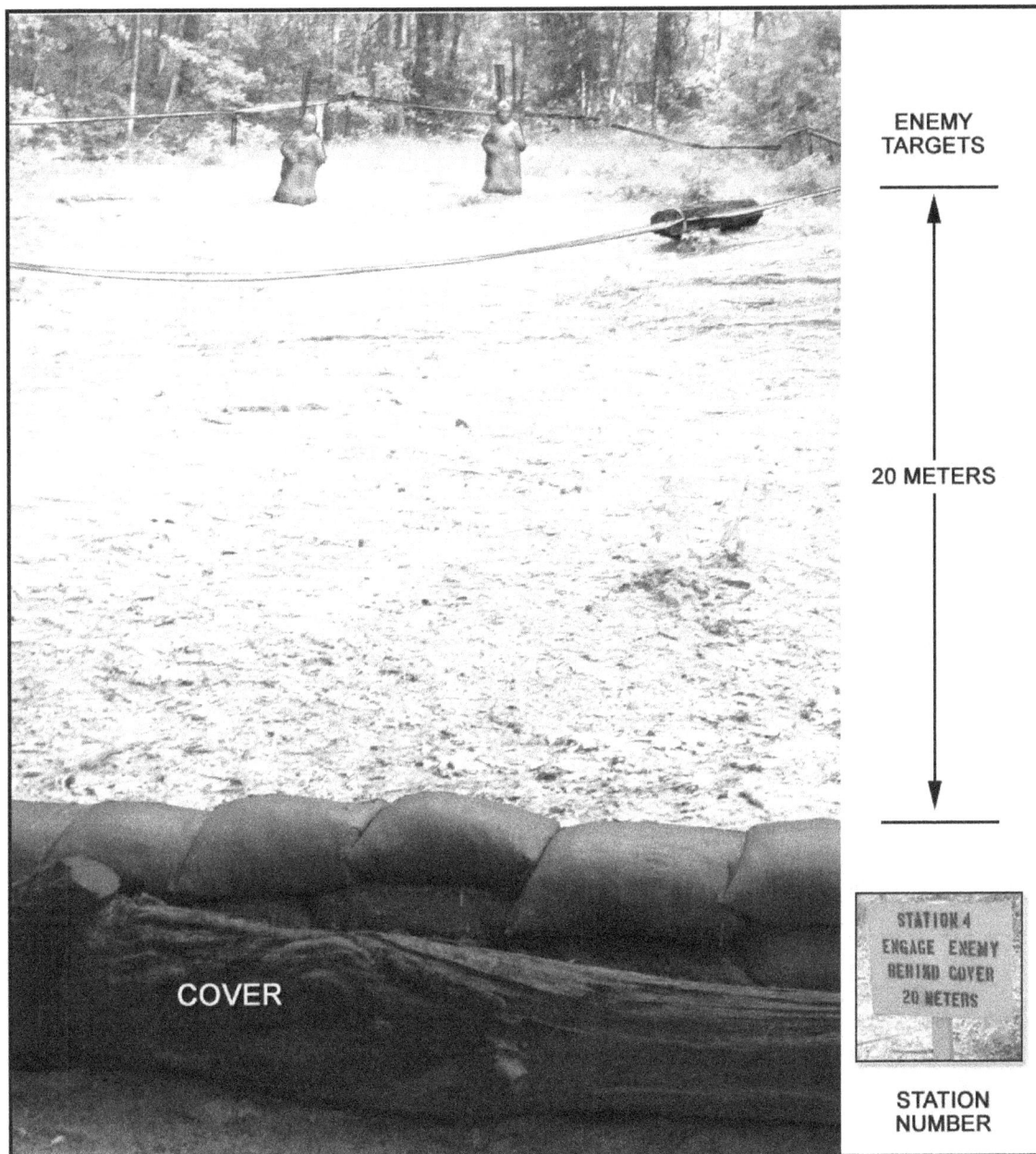

Figure 2-14. Station 4, engage enemy from behind cover (prone).

Figure 2-15. Station 5, engage trench (standing).

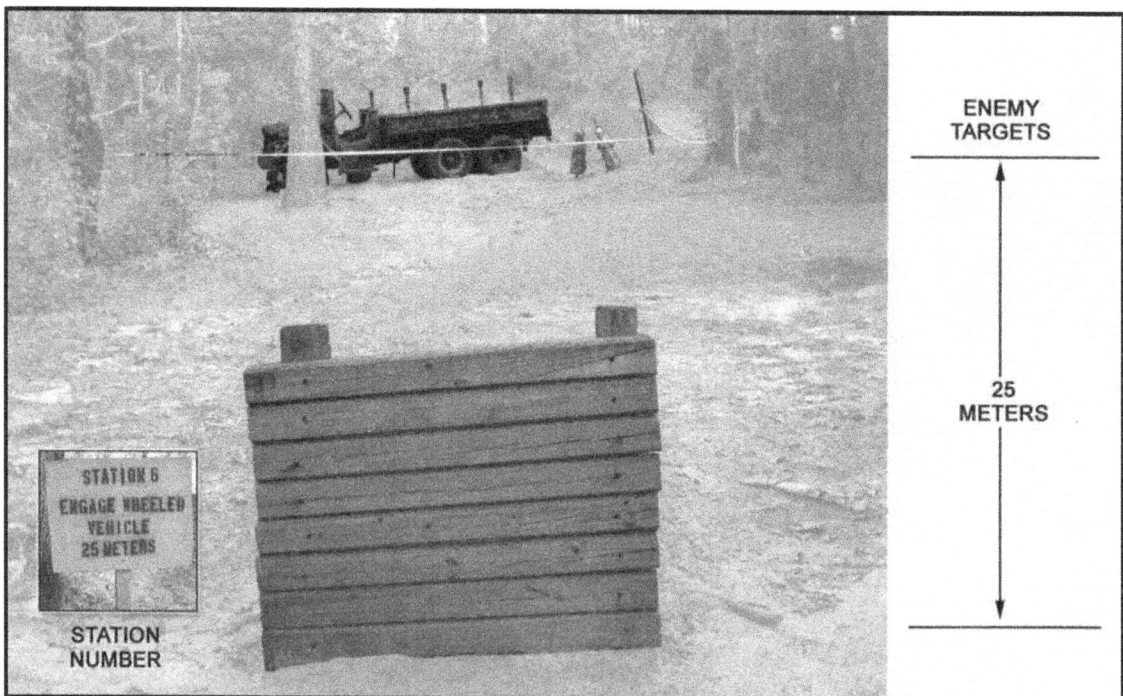

Figure 2-16. Station 6, engage wheeled vehicle (kneeling).

Figure 2-17. Station 7, identify hand grenades and pyrotechnic signals.

EQUIPMENT

2-73. The following is a minimum amount of range material and supplies needed to operate a practice and live hand grenade range.

- A helmet, a body armor vest, load-carrying equipment/enhanced tactical load-bearing vest (LCE/ETLBV), and ear protection for all range personnel and Soldiers attending training.
- Appropriate publications pertaining to training (FMs, TMs, ARs, SOPs).
- Range flag.
- Communications equipment.
- Targets in accordance with this manual.
- Grenades (live/practice) and pyrotechnics, as needed.
- Training aids, as needed.
- Ambulance or required dedicated evacuation vehicle.

NOTE: The driver must have knowledge of the route to the hospital.

- Potable water.
- Qualification scorecards in accordance with this manual.

PERSONNEL

2-74. In accordance with DA Pam 385-63, the following safety personnel are required for hand grenade training (Table 2-1):

- OIC.
- RSO.

2-75. Safe and successful performance of training also requires experienced support personnel. Support personnel required for training include—

- Pit safety NCOs.
- Ammunition personnel.
- Tower operator.
- Guards, as required.

- Medical personnel.
- Truck driver, if applicable.

Table 2-1. Officer in charge/range safety officer requirements.

WEAPON SYSTEMS	PERSONNEL REQUIREMENTS	
	OIC	RSO
Practice hand grenade, firing devices, simulators, or trip flares	SSG	SSG
Chemical agents and smoke[1]	SSG	NONE
Live grenades	SFC	SSG
Live-fire exercises, using organic weapons (squad through company, battery, and troop)	SFC	SSG
Combined arms live-fire exercises (CALFEXs) using outside fire support (section, platoon, squad, company, battery, troop, battalion, and squadron or larger)[2]	SFC	SSG

[1] When chemical, biological, radiological, nuclear (CBRN) training is being conducted, the OIC/RSO must be CBRN-qualified.
[2] The OIC will be a field-grade officer for battalion and larger-size units. The RSO on CALFEXs will be of the ranks listed above based on the complexity of the exercise and number of participants (i.e., squad, section, platoon, company, troop, squadron, battalion, and larger).

NOTE: Ranks of other services, DA civilians, and contractors must be equivalent to US Army ranks.

NOTE: OICs and RSOs involved in serious range incidents may lose their certification if determined to be in violation of AR 385-63 or DA Pam 385-63. While an incident is under investigation, their certificate may be suspended for as long as deemed necessary or revoked by the installation commander.

Officer in Charge and Noncommissioned Officer in Charge

2-76. The OIC or NCOIC is responsible for the overall conduct of the training, range orientation, range safety briefing, and briefing unit leaders. The OIC must have satisfactorily completed a standard program of instruction in the duties of the OIC (developed by the unit to which he is assigned) and attended a range safety briefing conducted by the installation range control. The OIC or NCOIC must—

- Be an E7 or above (NCOIC only).
- Be knowledgeable in the weapon systems involved and the duties required.
- Be certified by the commander.
- Receive instruction by range control.
- Have current safety cards.

NOTE: The rank of the OIC is determined by unit polices and regulations.

2-77. Once selected by the commander, the OIC should select the right personnel to conduct the training. Next, he should appoint an NCOIC who has current experience in the use of grenades and pyrotechnic signals. Together, the OIC and NCOIC should coordinate with adjacent units that are conducting or have conducted live grenade training for key personnel train-up and certification. The OIC and NCOIC should—

- Select and brief range support personnel on expected duties.
- Schedule for range certification with range control. If currently certified, review installation range instructions.
- Certify selected range personnel on their range duties.

NOTE: Before conducting training, the OIC and NCOIC should review FM 3-23.30, TM 9-1330-200-12, TM 9-1370-206-10, unit SOPs, AR 385-63, and DA Pam 385-63.

Range Safety Officer

2-78. The RSO should be the senior hand grenade instructor. The RSO must have satisfactorily completed a standard program of instruction in the duties of RSO (developed by the unit to which he is assigned) and attended a range safety briefing conducted by the installation range control. The RSO must—

- Be an E6 or above.
- Be knowledgeable in the weapon systems involved and the duties required.
- Ensure that the OIC has current safety cards.
- Perform no duties other than those of safety officer.

Pit Safety Noncommissioned Officers

2-79. Pit safety NCOs provide instruction, prepare practice grenades, and conduct practice and live hand grenade training safety. Range safety pit NCOs should—

- Be an E5 or above.
- Be knowledgeable in the weapon systems involved and the duties required.
- Be selected and certified on all hand grenade and pyrotechnic tasks by the OIC and NCOIC.

NOTE: These personnel require no safety cards, but must be task-certified by their unit on all grenade and pyrotechnic signals.

2-80. Pit safety NCOs also performs the five drop procedures in case of an emergency. These drop procedures are contingent on Soldier actions:

- Soldier milks the hand grenade.
- Soldier freezes after arming the hand grenade.
- Soldier remains standing after he throws the hand grenade downrange (attempting to observe the impact).
- Soldier drops the hand grenade.
- Soldier fails to take commands from the pit safety NCO.

Soldier Milks the Hand Grenade

2-81. Soldiers can milk the hand grenade in two ways; they can move their fingers or their thumb.

Moving the Fingers

2-82. When a Soldier milks the hand grenade by moving his fingers, the pit safety NCO performs the following procedures:

(1) Tell the Soldier to cease all action (**"FREEZE"**). The Soldier will close his hand.
(2) Decide if the grenade is armed or safe.
(3) Tell the Soldier to **"THROW"** if you determine that the grenade is armed or unsafe.

Moving the Thumb

2-83. When a Soldier milks the hand grenade by moving his thumb, the pit safety NCO performs the following procedures:

(1) Tell the Soldier to **"THROW"**.

NOTE: If the Soldier does not throw, perform the following actions.

(2) Place your thumb across the Soldier's thumb and your fingers across your Soldier's fingers, securing the Soldier's throwing hand (right hand for right-handed Soldiers, left hand for left-handed Soldiers).
(3) Use your free hand to force the Soldier to the front wall.

(4) Force the Soldier to drop the grenade over the front wall.

(5) Pull the Soldier into the pit, and protect him from the blast.

Soldier Freezes After Arming the Hand Grenade

2-84. When a Soldier freezes after arming the hand grenade, the pit safety NCO performs the following procedures:

(1) Tell the Soldier to **"THROW"**.

NOTE: If the Soldier does not throw, perform the following actions.

---

(2) Place your thumb across the Soldier's thumb and your fingers across your Soldier's fingers, securing the Soldier's throwing hand (right hand for right-handed Soldiers, left hand for left-handed Soldiers).

(3) Use your free hand to force the Soldier to the front wall.

(4) Force the Soldier to drop the grenade over the front wall.

(5) Pull the Soldier into the pit, and protect him from the blast.

Soldier Remains Standing After He Throws the Hand Grenade Downrange

2-85. Once the pit safety NCO sees the hand grenade leave the throwing pit, he ensures that the Soldier is kneeling. If the Soldier continues to stand, the pit safety NCO physically forces the Soldier to kneel.

Soldier Drops the Hand Grenade

2-86. If a live hand grenade is dropped accidentally after the safety pin has been removed, the pit safety NCO is responsible for reacting accordingly. He is responsible for the safety of the thrower and must decide what actions are the most appropriate. The pit safety NCO's actions are dependent upon the following factors:

- The location of the dropped grenade.
- The location of the thrower.
- The ability to physically move the thrower out of harm's way.

NOTE: All of these factors need to be considered before the safety pin is pulled.

2-87. Often, a pit safety NCO's actions depend upon the location of the designated safe area (dependant on the presence or absence of a knee wall).

Throwing Pit with Knee Wall

2-88. Knee walls provide a fast and safe means of reacting to a dropped grenade. In most instances, the pit safety NCO reacts to a dropped live grenade using the following procedures:

(1) Yell **"GRENADE"** to alert all other personnel in the area of the dropped grenade.

(2) Physically push the thrower over the knee wall.

(3) Fall on top of the thrower.

2-89. If a hand grenade is dropped over the knee wall, the pit safety NCO uses the following procedures:

(1) Yell **"GRENADE"**.

(2) Force the thrower to the ground inside the throwing pit.

Throwing Pit without Knee Wall

2-90. When using throwing pits that do not have knee walls, the pit safety NCO reacts to a dropped live grenade using the following procedures:

(1) Yell **"GRENADE"** to alert other personnel in the area.

(2) Physically move the thrower out of the throwing pit and into a safety pit.

2-91. If the hand grenade is dropped to the rear of the throwing pit, the pit safety NCO uses the following procedures:

(1) Yell **"GRENADE"**.

(2) Force the thrower over the front of the throwing pit.

(3) Follow the thrower.

WARNING

Do not kick or throw grenades into sumps. In response to a dropped grenade, Soldiers must move immediately from the danger area and drop to the prone position with the Kevlar helmet facing the direction of the grenade. This reduces the Soldier's exposure.

Soldier Fails to Take Commands

2-92. When a Soldier fails to take commands from the pit safety NCO, the pit safety NCO performs the following procedures:

(1) Repeat the command of prepare to throw.

NOTE: If the Soldier does not make an attempt to arm the grenade, perform the following actions.

(2) Place your hand over the Soldier's throwing hand (left hand for right-handed Soldiers, right hand for left-handed Soldiers), covering the fuze head.

(3) Place your other hand on the Soldier's helmet, and tell him to kneel in the throwing pit.

(4) Explain to the Soldier what he must do to throw the grenade downrange.

Ammunition Personnel

2-93. The ammunition personnel are in charge of accountability and handing out grenades.

NOTE: The ammunition NCO must attend an ammunition handler's class provided by the local ammunition supply point (ASP).

Tower Operator

2-94. The tower operator controls Soldier movements during range operations and maintains communications with range control.

Guards

2-95. Guards control traffic during range operations.

Medical Personnel

2-96. Medical support (with required medical supplies) must be present before and during range operations.

Truck Driver

2-97. The truck driver transports personnel to and from the range and provides support as needed (e.g., water, food, guard, etc.).

SURFACE DANGER ZONE

> **WARNING**
>
> Observe caution when using hand grenades or pyrotechnic signals with igniting type fuzes. These grenades and pyrotechnic signals ignite with a flash and should be thrown at least 10 meters from all friendly personnel to avoid hazardous conditions.

2-98. The surface danger zone (SDZ, shown in Figure 2-18) should be clear of all nonessential personnel before conducting LFXs. The impact area should also be level and free of debris.

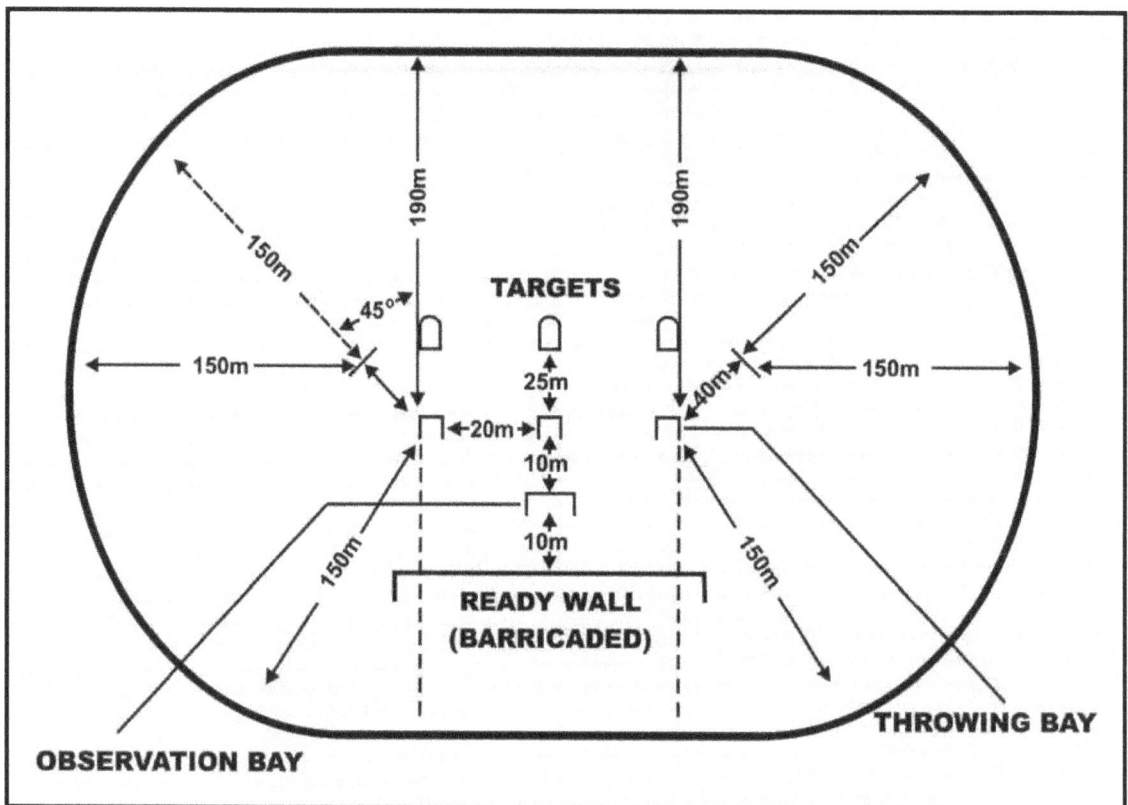

Figure 2-18. Surface danger zone for live-bay.

DUDS

2-99. Soldiers should treat any thrown grenade that fails to detonate as a dud, regardless of safety pin, safety clip, or safety lever status. Duds must be regarded as dangerous. During training, the pit safety NCO determines a dropped grenade's status (Table 2-2).

Table 2-2. Status of dropped grenade.

STATUS	DESCRIPTION
SAFE	A grenade with all safety devices intact
LIVE	A thrown grenade from the instant it is released until the expected fuze time has elapsed
DUD	A thrown grenade that failed to detonate after the expected fuze time has elapsed

---

CAUTION

Do not handle, approach, recover, or otherwise tamper with dud grenades. Let explosive ordnance disposal (EOD) personnel handle dud grenades.

Practice Grenade

2-100.   If a practice grenade does not detonate—
   (1)   Wait 5 minutes before defuzing the M69 practice grenade.
   (2)   Keep the bottom of the grenade oriented away from your body and pointed directly at the ground.
   (3)   Place the dud fuze in a sand-filled container, and return it to the issuing facility or dispose of it in accordance with unit SOP.

Fragmentation Grenade

2-101.   If a fragmentation grenade does not detonate in training—
   (1)   The thrower and pit safety NCO wait in the throwing pit for 5 minutes before returning to a covered area.
   (2)   The OIC or NCOIC notifies EOD immediately.

2-102.   No other grenades should be thrown in the area of the dud until it has been neutralized.

NOTE: If range facilities provide, continue training on an adjacent impact area that is separated by protective berms.

SECTION IV. TRAINING CONDUCT

Training conduct involves four steps:
   (1)   Occupy, inspect, and set up the range.
   (2)   Prepare for training.
   (3)   Conduct the training.
   (4)   Complete the training mission.

OCCUPY, INSPECT, AND SET UP RANGE

2-103.   The OIC must establish communication with the installation's range control and request permission to occupy the range before personnel, materiel, or supplies arrive. Once this has been accomplished, the OIC and NCOIC should—
   ●   Set up ammunition points and post guards.
   ●   Establish locations for medical station.
   ●   Designate Soldier holding areas.

- Establish water points.
- Designate parking areas.
- Inspect the range for operational conditions.
  - Check all throwing pits for sharp edges or unleveled throwing surfaces.
  - Ensure throwing pits meet standards.
  - Check the tower and tower public address (PA) system, if applicable.
- Request an opening code from range control, if applicable.
- Raise the range flag.

PREPARATION FOR TRAINING

2-104. The OIC and NCOIC should greet unit leaders and Soldiers as they arrive and direct them to the holding area. Actions at the holding area include—

- Ensuring all Soldiers attending training have a helmet, a body armor vest, LCE/ETLBV, ear protection, and a protective mask.
- Identifying Soldiers to be trained.
- Conducting a safety briefing (to include administrative personnel).

CONDUCT THE TRAINING

NOTE: The OIC should monitor all training activities.

2-105. There are two types of training: initial training for introducing Soldiers to hand grenades and pyrotechnic signals, and sustainment training of learned skills. The level of instruction will dictate the length and pace of training.

INITIAL TRAINING

2-106. Initial training involves the completion of six tasks:
- Participate in initial hand grenade training.
- Participate in distance and accuracy training.
- Participate in mock-bay training.
- Participate in live-bay training.
- Complete the hand grenade qualification course.

Participate in Initial Hand Grenade Training

2-107. The safety NCOs should take charge of the Soldiers to be trained and move them to the practice grenade training site. During the initial hand grenade training, the safety NCOs will demonstrate the proper techniques for employing hand grenades. Training includes—

- Hand grenade safety, inspection, and maintenance procedures.
- Proper hand grenade storage on Soldier equipment.
- Proper hand grenade throwing grip (left/right-hand grips).
- Proper hand grenade safety device removal.
- Proper hand grenade throwing positions and techniques.
- Demonstrations of pyrotechnic signals.

Participate in Distance and Accuracy Training

```
WARNING

Do not use live grenades for practicing distance and accuracy.
```

NOTES: 1. During the practice events and for qualification, each Soldier is required to throw several M69 practice hand grenades armed with the M228 detonating fuze. Although it takes only about a minute or less to install or replace a used M228 fuze, a company-size element will use several hundred. Preparing practice grenades for all participants is not feasible. Therefore, Soldiers should be given instruction on installing and removing a fired M228 fuze. See Appendix A for more information.

2. A confidence clip has been introduced to hand grenades and pyrotechnics equipped with a safety pin and pull ring. All hand grenades and the M228 fuze for the M69 practice grenade will come with the confidence clip attached. However, the confidence clip can be issued separately and then installed on the M228 fuze before connecting the fuze to the M69 practice grenade. See Appendix A for more information.

2-108. The distance and accuracy course is designed to develop the Soldier's proficiency in gripping and throwing hand grenades. When conducting the training, instructors should clarify the task, conditions, and standards for the course (Table 2-3), and demonstrate the task.

Table 2-3. Distance and accuracy course—task, condition, and standard.

TASK	Engage a variety of targets at varying ranges up to 40 meters.
CONDITION	Given 10 practice grenades, individual equipment, and a four-station course with a variety of targets at distances of 20, 30, and 40 meters.
STANDARD	The Soldier must successfully engage targets at each station with two out of three grenades. The Soldier must throw from the alternate prone, prone-to-kneeling, and prone-to-standing positions. A target is successfully engaged when the grenade detonates within 5 meters of the target.

2-109. To develop good safety habits, supervisors and instructors must ensure the Soldiers use proper throwing techniques.

NOTE: During the initial practical exercise, Soldiers should be allowed to observe the strike of the grenade so they can gain an appreciation for the weight of the grenade and the amount of force required to throw it accurately. After initial training, however, Soldiers should follow the proper procedures for seeking cover after throwing a grenade.

Participate in Mock-Bay Training

```
WARNING

Do not use live grenades for mock-bay training.
```

NOTE: Soldiers must attend mock-bay training before moving to live-bay training.

2-110.   Mock-bay training enables Soldiers to learn the proper techniques of throwing a hand grenade before moving to a live-bay, where Soldiers will experience realistic blast effects.

NOTE: Cadre/trainers use mock-bay training to identify Soldiers who require additional training and exhibit techniques that may be dangerous (to the Soldier and other personnel) in the live-bay environment.

Safety Precautions

2-111.   For the safe and effective conduct of mock-bay training, Soldiers must—
*   Wear LCE, body armor, ear protection, helmets, and eye protection, if available.
*   Receive a safety briefing before throwing training grenades.
*   Carry the hand grenades to the throwing pits using proper right- or left-hand grips.
*   Exhibit proper gripping procedures, throwing techniques, and throwing positions.
*   Be properly supervised by a mock-bay safety NCO.

Conduct

2-112.   At the mock-bay, cadre/trainers must—
*   Reinforce hand grenade safety.
*   Explain live-bay conduct.
    *   Safety precautions.
    *   Throwing order.
    *   Issue of hand grenades.
    *   Commands to prepare, throw, and take cover.
    *   Throwing position.

NOTE: Soldiers will throw from the standing position during mock-bay and live-bay training.

    *   Precautions taken in the event a Soldier drops a hand grenade after pulling the safety pin.

NOTE: The pit safety NCO is responsible for the safety of the thrower. The cadre/trainer will demonstrate the techniques he will use to protect the Soldier in the event a Soldier fails to take immediate cover behind the knee wall.

    *   Precautions taken in the event a Soldier refuses to throw a live hand grenade.

NOTE: If a Soldier refuses to throw a live hand grenade while in the live-bay pit, the pit safety NCO directs the Soldier to remain behind cover and throws the grenades. The Soldier is then removed from the remainder of training, and the Soldier's name is given to the chain of command.

2-113.   Then, training conduct proceeds as follows:
(1) Each Soldier is issued four M69 hand grenades with the M228 fuze. The Soldiers install the M228 fuze prior to entering the mock-bay.
(2) Soldiers are rotated twice through mock-bay:
    *   The first rotation is for observing Soldier actions and correcting throwing deficiencies.

NOTE: Soldiers are not allowed to stand and observe a thrown hand grenade. They are directed to drop to a kneeling position behind cover after throwing the second hand grenade.

■ The second rotation ensures Soldiers are performing as taught and serves to identify Soldiers who demonstrate weak or dangerous technique.

NOTE: Soldiers identified as high-risk should be tagged (with something that can be seen from a distance) and sent back for reinforcement training. These Soldiers should be placed at the end of the throwing order to ensure that training is not hindered and to minimize exposing other Soldiers to potential risks.

Participate in Live-Bay Training

NOTE: Soldiers going to the live-bay must have first practiced all the procedures in the mock-bay.

2-114. After all grenades have been inspected, the NCOIC should take charge and move the Soldiers to the live-bay throwing site. Live-bay training gives Soldiers the opportunity to experience throwing a live fragmentation hand grenade.

Safety Precautions

2-115. Throwing of live hand grenades can be done in a safe manner if the range safety procedures are followed. These procedures include identification of high-risk Soldiers who had problems throwing grenades during the initial training block of instruction.

NOTE: Live-bay range personnel must be completely alert at all times and prepared to take appropriate actions for any given situation. The range OIC and NCOIC must be positioned to observe the throw phase and count grenade explosions for purposes of grenade accountability and duds.

2-116. For the safe conduct of live-bay training, Soldiers must—
- Complete mock-bay training before throwing live grenades.
- Wear LCE, body armor, ear protection, helmets, and eye protection, if available.
- Receive a safety briefing before throwing live grenades.
- Carry the hand grenades to the throwing pits using proper right- or left-hand grips.
- Be properly supervised by a live-bay safety NCO.
- Be behind protective barriers, where they will stay until called forward.

CAUTION

All Soldiers, including range personnel, and visitors to the range must be behind protective barriers and wear appropriate safety gear.

2-117. The OIC and NCOIC must take the appropriate safety precautions before conduct of live-bay training. This includes—

- Ensure all guards are posted and roadblock barriers are in place before moving Soldiers to the live-bay.
- Ensure communication between roadblocks, the RSO, and the tower is confirmed before live throwing.
- Ensure Soldiers are shown the live-bay training area and a safety briefing is given on the operating procedures of the live-bay.

NOTE: Range safety personnel must reinforce the fact that cooking off hand grenades is not allowed in the live-bay. Soldiers must also be briefed on the actions to take for a dropped grenade.

- Request an opening code to go HOT.

Conduct

2-118. During conduct of live-bay training—

(1) After the live-bay safety briefing, the RSO moves the Soldiers to the ready line.
(2) The ammunition NCO issues grenades to the Soldiers, making sure they are holding the grenades properly.

NOTE: Grenades are issued only to those Soldiers who are in line to go to the throwing pit. Pit safety NCOs will make sure the Soldiers are holding the grenades properly and at the chin/chest level. The pit safety NCOs will direct Soldiers to sound off, each indicating with which hand he will throw the grenade.

(3) The pit safety NCOs will direct Soldiers to specific throwing pits.

NOTE: A pit safety NCO is assigned to each throwing pit.

(4) The pit safety NCO observes the movement of Soldiers to the pit.
(5) The Soldier going into the live-bay identifies the throwing arm to the pit safety NCO. The pit safety NCO directs the Soldier to the appropriate position of the pit, based on left or right throwing arm. The pit safety NCO directs the Soldier to hand over the grenade in his nonthrowing hand.
(6) The pit safety NCO directs the Soldier to remove the safety clip and safety pin, and to prepare to throw.

NOTE: From this point on, the pit safety NCO does not divert his eyes from the throwing hand until completion of the throw.

WARNING

If a grenade is dropped in the pit, the pit safety NCO forces the Soldier out of the pit and into the designated safe area and follows him.

---

(7) The pit safety NCO signals the tower that the Soldier is prepared to throw by holding up his left or right arm.
(8) When all throwing pits are ready, the tower operator commands **"THROW"**, and the pit safety NCO repeats the **"THROW"** command to the Soldier in the pit.

NOTE: All pits throw at the same time.

> **WARNING**
>
> **If a Soldier releases the safety lever but fails to throw the grenade, the pit safety NCO forcefully repeats the command to throw; if necessary, the pit safety NCO grabs and throws the grenade himself.**

(9) The Soldier throws the grenade, and then drops to cover.

> **WARNING**
>
> **If a Soldier does not take cover, the pit NCO forces the Soldier to take cover.**

(10) The tower operator commands **"CLEAR"** after observing each grenade detonation. The pit safety NCO hands the thrower the second grenade.

(11) The Soldiers prepare to throw a second grenade, repeating the required steps.

Complete the Hand Grenade Qualification Course

NOTE: This course should not be attempted until all initial training has been completed.

2-119.   The purpose of the qualification course is to measure and evaluate the Soldier's ability to engage a variety of targets in natural terrain under simulated combat conditions. The qualification course allows the Soldier to gain confidence in arming and throwing hand grenades in a simulated tactical scenario.

Stations

NOTE: Active duty Soldiers in Infantry and reconnaissance units must qualify on the hand grenade qualification course every six months. All other active duty Soldiers and those in Army Reserve and National Guard units must qualify once a year.

2-120.   The hand grenade qualification course (Figures 2-10 through 2-17) is standardized throughout the Army. It consists of seven stations (Table 2-4), with a minimum of one grader at each station. The course is conducted in two-man teams, but Soldiers are evaluated individually. Each participant is issued ten hand grenades and must successfully engage seven targets.

NOTE: See TC 25-8 for more information about the hand grenade qualification course.

Table 2-4. Hand grenade qualification course stations.

STATION	TASK	CONDITION	STANDARD
1	Engage a group of F-type silhouette targets in the open from a two-man fighting position.	The targets are located 35 meters to the front of the fighting position, simulating enemy movement through and beyond the squad's protective wire.	No more than two grenades should be used on any target. Only one is used if the first grenade is on target.
2	Engage a bunker using available cover and concealment.	The bunker can have one or two firing portholes oriented toward the direction of the buddy team's movement and a rear exit.	
3	Engage a fortified enemy mortar position.	The fortified enemy mortar position must be located 20 meters away.	
4	Engage a group of enemy targets.	The group of enemy targets must be behind cover and located 20 meters away.	
5	Clear an entry point to a trench line.	The trench line must be located 25 meters away.	
6	Engage enemy troops in a halted, open-type wheeled vehicle.	The halted, open-type wheeled vehicle must be located 25 meters away.	
7	Identify hand grenades and pyrotechnic signals.	All grenades must present proper shape, color, and markings.	Soldiers must be able to identify grenades and pyrotechnic signals by shape, color, markings, and capabilities.

Scoring

2-121. Although no two hand grenade qualification courses are alike, the standards must be consistent. Qualification must be awarded only to those Soldiers who meet these standards. The minimum course standards should include live-bay training and the hand grenade qualification course.

NOTE: The evaluator at each station determines scoring in accordance with DA Form 3517-R (Hand Grenade Qualification Scorecard, shown in Figures 2-19 and 2-20).

HAND GRENADE QUALIFICATION SCORECARD
For use of this form, see FM 3-23.30; the proponent agency is TRADOC

NOTE: In addition to the requirements on the scorecard, the Soldier must throw two live fragmentation grenades to qualify.

A. NAME (Last, First, Middle Initial)	B. UNIT		C. DATE (YYYYMMDD)
Bradshaw, PFC Zane F.	A 2/29		20091120

D. EVALUATOR'S NAME	E. DATE LIVE GRENADES WERE THROWN	(YYYYMMDD)
Duffell, SFC Jerry		20091120

F. STATION	G. TYPE TARGET	H. GO	I. NO-GO	J. EVALUATOR'S INITIALS
1	Engage Enemy from Fighting Position at 35 Meters (Standing)	X	☐	
2	Engage Bunker (Prone)	X	☐	
3	Engage Enemy Mortar Position at 25 Meters (Kneeling)	☐	X	
4	Engage Enemy Behind Cover at 20 Meters (Prone)	X		
5	Engage Trench at 25 Meters (Standing)	X	☐	
6	Engage Wheeled Vehicle at 25 Meters (Kneeling)	X	☐	
7	Identify Hand Grenades and Pyrotechnic Signals	X	☐	

K. QUALIFICATION STANDARD		CHECK ONE
PASSED 7	EXPERT	☐
PASSED 6	FIRST CLASS	X
PASSED 5	SECOND CLASS	☐
PASSED 4 OR LESS	UNQUALIFIED	☐

L. DATE INITIALED (YYYYMMDD)	M. SCORER'S INITIALS
20091120	NAH

N. DATE INITIALED (YYYYMMDD)	O. SCORER'S INITIALS
20091120	JEM

DA FORM 3517-R, NOV 2009   PREVIOUS EDITION IS OBSOLETE.   APD PE v1.00

EXAMPLE

**Figure 2-19. Example of a completed DA Form 3517-R
(Hand Grenade Qualification Scorecard) (front).**

PERFORMANCE MEASURES	GO	NO-GO	PERFORMANCE MEASURES	GO	NO-GO
STATION 1. Engage Enemy From Fighting Position at 35 Meters *(Standing)*			**STATION 5. Engage Trench at 25 Meters** *(Standing)*		
A. Detonated at least one grenade within 5 meters of the center of target.	X		A. Detonated at least one grenade inside trench.	X	
B. Kept exposure time under 3 seconds.	X		B. Kept exposure time under 3 seconds.	X	
C. Returned to covered position after each throw.	X		C. Returned to covered position after each throw.	X	
D. Used proper grip.	X		D. Used proper grip.	X	
E. Used proper throwing techniques.	X		E. Used proper throwing techniques.	X	
F. Completed performance measures 1A through 1E within 15 seconds.	X		F. Completed performance measures 5A through 5E within 15 seconds.	X	
STATION 2. Engage Bunker *(Prone)*			**STATION 6. Engage Wheeled Vehicle at 25 Meters** *(Kneeling)*		
A. Approached from blind side.		X	A. Detonated within 1 meter of vehicle or within 5 meters of dismounting troops.	X	
B. Checked for bunker opening.	X		B. Kept exposure time under 3 seconds.	X	
C. Detonated grenade in bunker.	X		C. Returned to covered position after each throw.	X	
D. Rolled away from bunker.	X		D. Used proper grip.	X	
E. Used proper grip.	X		E. Used proper throwing techniques.	X	
F. Used cook-off technique.	X		F. Completed performance measures 6A through 6E within 15 seconds.	X	
G. Completed performance measures 2A through 2F within 15 seconds.	X				
STATION 3. Engage Mortar Position at 25 Meters *(Kneeling)*			**STATION 7. Identify Hand Grenades and Pyrotechnic Signals**		
A. Detonated at least one grenade inside mortar position.	X		A. Selected fragmentation grenade to engage enemy soldiers.	X	
B. Kept exposure time under 3 seconds.	X		B. Identified M83 grenade as "White Smoke" or "HC Smoke."	X	
C. Returned to covered position after each throw.	X		C. Identified M18 grenades as "Colored Smoke" or "Purple *(and so forth)* Smoke." *(If specific color is stated, it must be the same as color on the training aid grenade used.)*	X	
D. Used proper grip.	X		D. Identified M7A2/A3 grenade as CS or riot control.	X	
E. Used proper throwing techniques.	X		E. Identified M14 grenades as incendiary.	X	
F. Completed performance measures 3A through 3E within 15 seconds.	X				
STATION 4. Engage Enemy Behind Cover at 20 Meters *(Prone)*			**NOTES:**		
A. Detonated at least one grenade within 5 meters of the center of target.	X		1. FOR PERFORMANCE MEASURES 7A THROUGH 7E, IF THE EXAMINEE CANNOT CORRECTLY STATE THE NAME OF THE GRENADE, BUT CAN CORRECTLY IDENTIFY ITS USE, THEN THE EXAMINEE WILL BE SCORED A "GO."		
B. Kept exposure time under 3 seconds.	X				
C. Returned to covered position after each throw.	X		2. EACH PERFORMANCE MEASURE AT EACH SECTION IS GRADED ON A PASS/FAIL STANDARD. A SOLDIER MUST PASS ALL OF THE STANDARDS TO RECEIVE A "GO" ON THAT STATION.		
D. Used proper grip.	X				
E. Used proper throwing techniques.	X				
F. Completed performance measures 4A through 4E within 15 seconds.	X				

Page 2, DA Form 3517-R, NOV 2009   APD PE v1.00

Figure 2-20. Example of a completed DA Form 3517-R (Hand Grenade Qualification Scorecard) (back).

SUSTAINMENT TRAINING

2-122. Units should integrate the use of grenades into collective tasks, rather than training these skills as a separate event. Leaders at all levels should study the employment of grenades in conjunction with the unit mission and implement a training program that supports that mission.

2-123. The following generic tasks can assist units in training and evaluating hand grenade proficiency:

- Identify hand grenades and pyrotechnic signals.
- Inspect and maintain hand grenades and pyrotechnic signals.
- Employ a pyrotechnic signal.
- Complete the distance and accuracy course.
- Complete the bunker complex course.
- Complete the trench complex course.
- Complete the building complex course.
- Participate in mock-bay training.

Identify Hand Grenades and Pyrotechnic Signals

2-124. The purpose of this training is to develop and—at a later time—test the Soldier's knowledge in identifying and explaining the use of hand grenades and pyrotechnic signals. When conducting the training, instructors should clarify the task, conditions, and standards (Table 2-5).

Table 2-5. Identify hand grenade and pyrotechnic signals—task, condition, and standard.

TASK	Identify a variety of hand grenades and pyrotechnic signals.
CONDITION	Given individual equipment, and a variety of hand grenade and pyrotechnic signal pictures or mock-up training aids, and a mix of 3x5 cards with the name of each grenade and pyrotechnic signal.
STANDARD	The Soldier must successfully align the correct 3x5 card with the picture or mockup training aid of a grenade and pyrotechnic written on the card, and explain the characteristics, capability, and purpose of each.

Inspect and Maintain Hand Grenades and Pyrotechnic Signals

2-125. Hand grenades and pyrotechnic signals must be inspected on a regular basis to ensure serviceability due to prolonged exposure to environmental conditions, damage in shipping or in storage, or missing safety devices. Maintenance requirements are minimal (Table 2-6).

NOTE: See TM 9-1330-200-12 and TM 9-1370-206-10 for more information.

Table 2-6. Inspect and maintain hand grenades and pyrotechnic signals— task, condition, and standard.

TASK	Perform preventive maintenance checks and services (PMCS) of hand grenades and pyrotechnic signals.
CONDITION	Given individual protective equipment; a clean lint free cloth; an oil-free camel hair brush; cleaner, lubricant, and preservative (CLP); and an M67 or M69 hand grenade, a smoke grenade, and a signal flare.
STANDARD	The Soldier must inspect, clean, and store M67 or M69 hand grenades and pyrotechnic signals without causing damage to the hand grenade or pyrotechnic signal or cause injury to friendly personnel, in accordance with TM 9-1330-200-12 and TM 9-1370-206-10.

Employ a Pyrotechnic Signal

2-126. To successfully employ pyrotechnic signals, Soldiers must (Table 2-7)—
- Be familiar with the type of pyrotechnic signal used.
- Know how to properly hold, prepare, and throw or launch the pyrotechnic signal from various positions.

NOTE: See Chapter 4 for employment and safety considerations.

Table 2-7. Employ a pyrotechnic signal—task, condition, and standard.

TASK	Employ a pyrotechnic signal.
CONDITION	Given individual protective equipment, an M106 and M18 smoke grenade, a handheld flare, and a 20-meter target.
STANDARD	The Soldier must provide a complete smoke screen that cannot be observed from the target area, and successfully launch a handheld flare (by hand or from the ground) to communicate withdrawal or shifting of friendly fires.

Complete the Distance and Accuracy Course

WARNING

Do not use live grenades for practicing distance and accuracy.

2-127. The distance and accuracy course is designed to develop the Soldier's proficiency in gripping and throwing hand grenades. When conducting the training, instructors should clarify the task, conditions, and standards for the course (Table 2-8), and demonstrate the task.

Table 2-8. Distance and accuracy course—task, condition, and standard.

TASK	Engage a variety of targets at varying ranges up to 40 meters.
CONDITION	Given 10 practice grenades, individual equipment, and a four-station course with a variety of targets at distances of 20, 30, and 40 meters.
STANDARD	The Soldier must successfully engage targets at each station with two out of three grenades. The Soldier must throw from the alternate prone, prone-to-kneeling, and prone-to-standing positions. A target is successfully engaged when the grenade detonates within 5 meters of the target.

2-128. To develop good safety habits, supervisors and instructors must ensure the Soldiers use proper throwing techniques.

NOTE: During the initial practical exercise, Soldiers should be allowed to observe the strike of the grenade so they can gain an appreciation for the weight of the grenade and the amount of force required to throw it accurately. After initial training, however, Soldiers should follow the proper procedures for seeking cover after throwing a grenade.

Complete the Bunker Complex Course

```
                         WARNING

When using M106 smoke grenades, DO NOT cook off the grenade.
These smoke grenades have a flash-bang grenade fuze with a 0.7-
to 2-second time delay.
```

```
                         WARNING

Do not use live grenades for practicing the bunker complex course.
```

2-129.   The bunker complex course exercise develops the Soldier's proficiency in properly attacking a bunker complex from a covered and concealed location while using obscuration and proper movement techniques. When conducting the training, instructors should clarify the task, conditions, and standards for the course (Table 2-9), and demonstrate the task.

NOTE: See Chapter 3 for proper cook-off technique.

Table 2-9. Bunker complex course—task, condition, and standard.

TASK	Engage an enemy bunker complex.
CONDITION	Given an individual weapon, helmet, LCE and body armor, cover and concealment, and two M69 hand grenades with M228 fuze, or M84 flash-bang grenades, and one M106 or M83 smoke grenade.
STANDARD	The Soldier must successfully engage and disable a bunker. The Soldier must provide a smoke screen (M106 or M83) to cover the approach to the bunker from the blind side, properly cook off a M69 grenade, put the grenade into the firing port of the bunker, roll away from the bunker, and turn 180 degrees to cover the rear exit of the bunker. The grenade must detonate in the bunker.

NOTE: See Chapter 3 for proper M69 cook-off technique.

Complete the Trench Complex Course

```
                         WARNING

When using M106 smoke grenades, DO NOT cook off the grenade.
These smoke grenades have a flash-bang grenade fuze with a 0.7-
to 2-second time delay.
```

```
┌────────────────────────────────────────────────────────────┐
│                                                            │
│                        WARNING                             │
│                                                            │
│   Do not use live grenades for practicing the trench complex │
│   course.                                                   │
│                                                            │
└────────────────────────────────────────────────────────────┘
```

2-130.   The trench complex course exercise develops the Soldier's proficiency on how to properly attack a trench complex from a covered and concealed location while using obscuration and proper movement techniques. When conducting the training, instructors should clarify the task, conditions, and standards for the course (Table 2-10), and demonstrate the task.

Table 2-10. Trench complex course—task, condition, and standard.

TASK	Engage an enemy trench complex.
CONDITION	Given an individual weapon, helmet, LCE and body armor, cover and concealment, and two M69 hand grenades with M228 fuze, and one M106 or M83 smoke grenade.
STANDARD	The Soldier must successfully enter a trench and engage enemy personnel. The Soldier must provide a smoke screen (M106 or M83) to cover the approach to the trench from the blind side, properly cook off a M69 grenade, throw or roll the grenade into the trench, roll away from the mouth of the trench, wait for the explosion, and enter and clear the trench. The grenade must detonate in the trench.

NOTE: See Chapter 3 for proper M69 cook-off technique.

Complete the Building Complex Course

2-131.   The building complex course exercise develops the Soldier's proficiency on how to properly attack and clear a building from a covered and concealed location while using obscuration and proper movement techniques. When conducting the training, instructors should clarify the task, conditions, and standards for the course (Table 2-11), and demonstrate the task.

```
┌────────────────────────────────────────────────────────────┐
│                                                            │
│                        WARNINGS                            │
│                                                            │
│   When using M84 grenades or other forms of flash-bang     │
│   grenades, DO NOT cook off the grenade. These grenades    │
│   have a 1- to 2.3-second time delay.                      │
│                                                            │
│   When using M106 smoke grenades, DO NOT cook off the      │
│   grenade. These smoke grenades have a flash-bang grenade  │
│   fuze with a 0.7- to 2-second time delay.                 │
│                                                            │
│   DO NOT use M18, M83, and AN-M8 HC smoke grenades in      │
│   enclosed or confined spaces. Burning-type grenades burn  │
│   oxygen. Standard protective masks filter particles but   │
│   will not supply oxygen.                                  │
│                                                            │
└────────────────────────────────────────────────────────────┘
```

CAUTION

M18, M83, and AN-M8 HC have the potential to start fires when thrown on dry tender.

Table 2-11. Building complex course—task, condition, and standard.

TASK	Enter and clear a building complex.
CONDITION	As a member of a clearing team. Given an individual weapon, helmet, LCE and body armor, cover and concealment, and two M69 hand grenades with M228 fuze, or M84 flash-bang grenades, and one M106 or M83 smoke grenade.
STANDARD	The Soldier must successfully enter a building and engage enemy personnel. The Soldier must provide a smoke screen (M106 or M83) to cover the approach to the building from a covered and concealed position, properly prepare the grenade (cook off a M69 or use a M84 flash-bang), throw the grenade into a designated building entry point, take cover, wait for the explosion, and enter and clear the entry point. The grenade must detonate in the building/room.

NOTE: See Chapter 3 for proper M69 cook-off technique.

Participate in Mock-Bay Training

WARNING

Do not use live grenades for mock-bay training.

2-132. In mock-bay training, Soldiers practice throwing grenades before moving to live-bay training. This training introduces the Soldier to throwing commands and provides additional throwing practice. Instructors should orient the Soldiers to the mock-bay training pit and explain the commands that are used during actual throwing. When conducting the training, instructors should clarify the task, conditions, and standards for the training (Table 2-12), and demonstrate the task. Soldiers should also practice the procedures used during live-bay training.

NOTE: The instructor must reinforce correct throwing and safety procedures.

Table 2-12. Mock-bay training—task, condition, and standard.

TASK	Engage targets in the open with hand grenades.
CONDITION	Given individual equipment, to include helmet, body armor, and two M69 hand grenades with M228 fuze, a mock-bay pit that replicates a live-bay pit, ear protection, and an orientation and safety briefing.
STANDARD	Soldiers must safely carry, arm, and throw two practice hand grenades from the mock-bay pit while following the commands from the instructor or NCOIC. Soldiers must not move from the cover of the pit until the command "CLEAR", "ALL CLEAR" is given.

NOTE: Be sure the physical layout of the mock-bay pit replicates the live-bay pit. This not only gives the Soldier the sensation of throwing a live fragmentation hand grenade, but also instills confidence in his ability to throw the hand grenade.

Participate in Live-Bay Training

> **NOTE:** Soldiers going to the live-bay must have first practiced all the procedures in the mock-bay.

2-133.   After all grenades have been inspected, the NCOIC should take charge and move the Soldiers to the live-bay throwing site. Live-bay training gives Soldiers the opportunity to experience throwing a live fragmentation hand grenade (Table 2-13).

Table 2-13. Live-bay training—task, condition, and standard.

TASK	Engage targets in the open with hand grenades.
CONDITION	Given individual equipment, to include helmet, body armor, and two M67 fragmentation hand grenades, a live-bay pit, ear protection, and an orientation and safety briefing.
STANDARD	Soldiers must safely carry, arm, and throw two M67 fragmentation hand grenades from the live-bay pit while following the commands from the instructor or NCOIC. Soldiers must not move from the cover of the pit until the command "CLEAR", "ALL CLEAR" is given.

> **NOTE:** For more information about live-bay training, see the initial training segment of this chapter.

COLLECTIVE TRAINING

2-134.   Once Soldiers can identify grenades and pyrotechnics, have an understanding of their purposes, and can safely arm and throw live hand grenades and pyrotechnics, units should integrate their use into collective tasks. The hand grenade qualification course is optional; commanders can modify the stations or design training that supports unit METLs.

> **NOTE:** Soldiers should use the M69 practice hand grenades against realistic targets while practicing the collective tasks. Use the M84 or other types of flash-bang hand grenades during urban operations, if available. However, the M69 can be used in lieu of the M84.

Squad Situational Training Exercise

2-135.   Leaders should present Soldiers with tactical situations in a realistic squad STX integrating the use of hand grenades and pyrotechnics with other fire team or squad weapons to force Soldiers to make sound tactical decisions and improve their skills.

2-136.   Leaders should consider integrating hand grenade tasks that are best suited to the unit's METL into the tactical scenario. Figure 2-21 is an example squad STX that tests the squad's weapon proficiency.

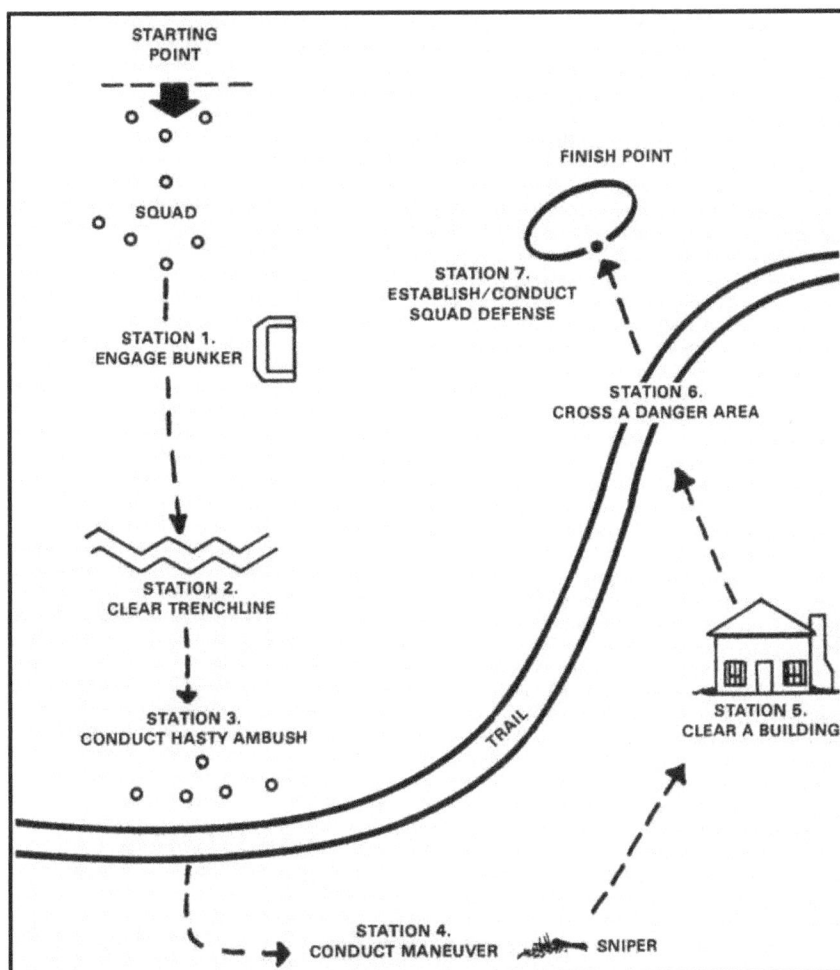

Figure 2-21. Example of squad situational training exercise with hand grenades.

COMPLETE THE TRAINING MISSION

2-137. At the completion of training, all equipment, range material, and ammunition should be accounted for, to include the completion of range maintenance, the OIC and RSO can close the range. This includes—

- Request a closing code from range control.
- Release unit Soldiers.
- Remove all equipment and ammunition from the range.

NOTE: Turn in all unexpended grenades in original grenade containers to the ASP, along with all safety pins and packing residue from all detonated grenades.

- Have EOD find and clear any duds or grenades thrown without the safety pin pulled.
- Police the range, fill in all holes with sand, rake the impact area, and perform other range maintenance as required by local SOP.
- Request a range inspection from range control when ready to clear.
- Turn in paperwork and equipment.
- Submit after action report to headquarters.
- Report any noted safety hazards to proper authorities.

This page intentionally left blank.

Chapter 3

Hand Grenades

Hand grenades are used to supplement small arms against an enemy in close combat, and for non-lethal operations to temporarily stun enemy personnel. Proper control and safety procedures allow for safe employment of hand grenades.

INSPECTION

WARNING

Hand grenades—like any other weapon—must be inspected before use and properly secured to avoid serious injury or death.

3-1. Soldiers should perform three types of inspections:
- Initial inspection.
- Before storing.
- Daily checks.

CAUTION

Observe precautions generally applicable to use of ammunition.

NOTE: See TM 9-1330-200-12 for more information about hand grenade inspection. These inspections focus on the M67 fragmentation grenade; however, all grenades with fuzes must be inspected.

INITIAL INSPECTION

3-2. When in bulk, hand grenades are secured in shipping containers (Figure 3-1). Personnel should inspect the shipping container upon receipt. Shipping containers that are damaged should not be opened; they should be returned to the ASP or disposed of using the methods outlined in the unit SOP.

Figure 3-1. Hand grenade shipping container.

3-3. Each grenade within the shipping container is housed in sealed individual canisters (Figure 3-2). Hand grenades may be issued while still in their individual shipping canisters or unpacked and issued by someone within the chain of command.

3-4. Upon removing the sealed individual canisters from the shipping container, personnel should inspect the canisters and identify any of the following discrepancies:

- The canister has been damaged.
- The seal on the canister has been tampered with or is missing.

Figure 3-2. Hand grenade shipping canister.

3-5. Personnel should open the canister. Once the canister has been opened but before removing the packing material (Figure 3-3), personnel should inspect the grenade and identify any of the following discrepancies:

- The hand grenade is upside down inside of the shipping canister.
- The safety pin is not properly attached or is missing.

Figure 3-3. Hand grenade shipping canister with packing material.

3-6.   Then, personnel should remove the packing material and the grenade from the canister. Once the packing material and the grenade has been removed from the canister (Figure 3-3), personnel should inspect the grenade and identify any of the following discrepancies:

- Rust is on the body or the fuze.
- Holes are visible in the body or the fuze.

NOTE: If any of the discrepancies are found upon receipt of newly issued hand grenades, personnel should return the grenade and shipping canister to the issuing person or dispose of it in accordance with the unit SOP.

BEFORE STORING

3-7.   Before securing the hand grenades in ammunition pouches, personnel should take the following safety precautions:

- Inspect grenades to ensure all safety devices are present (Figure 3-4):
  - Safety pin (1) with pull ring and confidence clip (2).
  - Safety clip (3).
  - Safety lever (4).
- Ensure all the safety devices are intact and serviceable.
- Check the grenade fuze for tightness. It must be tightly fitted within the grenade body (5).

WARNING

Never remove the fuze from a live grenade.

- Ensure the lever is not bent or broken.
- Check the body for rust or dirt.

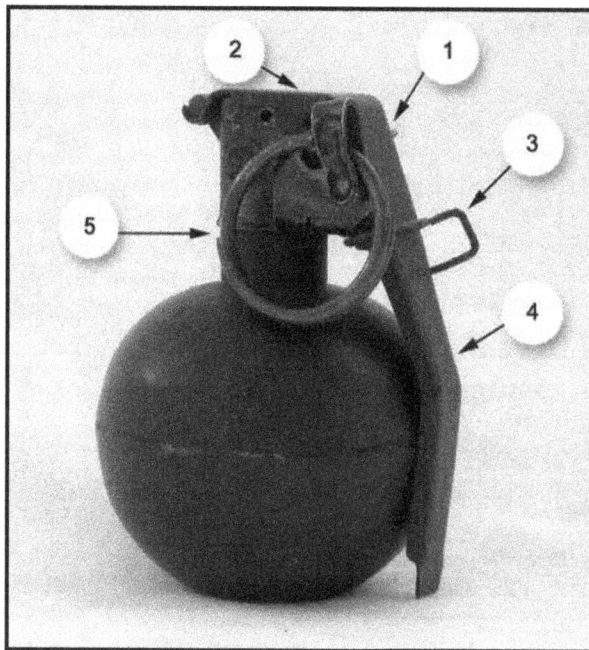

Figure 3-4. Hand grenade with safety clip installed.

Safety Clip Installation

NOTE: A safety clip can be removed and reattached to a hand grenade if the safety pin is still in place.

3-8. Hand grenades equipped with a safety clip, which prevents the fuze safety lever from springing loose, may come loose during shipping or when stowing or removing from the grenade carrying pouch. When installed correctly, the safety clip secures the grenade fuze lever even if the safety pin assembly is accidentally removed.

NOTE: Not all hand grenades have safety clips. See TM 9-1330-200-12 for more information about the types of grenades that do require and do not require a safety clip.

3-9. Once hand grenades have been removed from their shipping canister, personnel must ensure that the safety clip is present and install a safety clip, if required. The safety clip is adaptable to M26 and M67-series grenades, the MK2 grenade, and the M69 practice grenade.

NOTE: If lost or broken, safety clips for some types of grenades can be procured through Class V ammunition supply channels (NSN 1330-00-183-5996).

3-10. To install a safety clip (Figure 3-5)—

(1) Hold the fuzed grenade in the palm of your hand with the pull ring up.
(2) Insert the small loop at the open end of the safety clip into the slot of the fuze body beneath the safety lever.
(3) Press the clip across the safety lever until the closed end of the clip touches the safety lever and snaps securely into place around it.

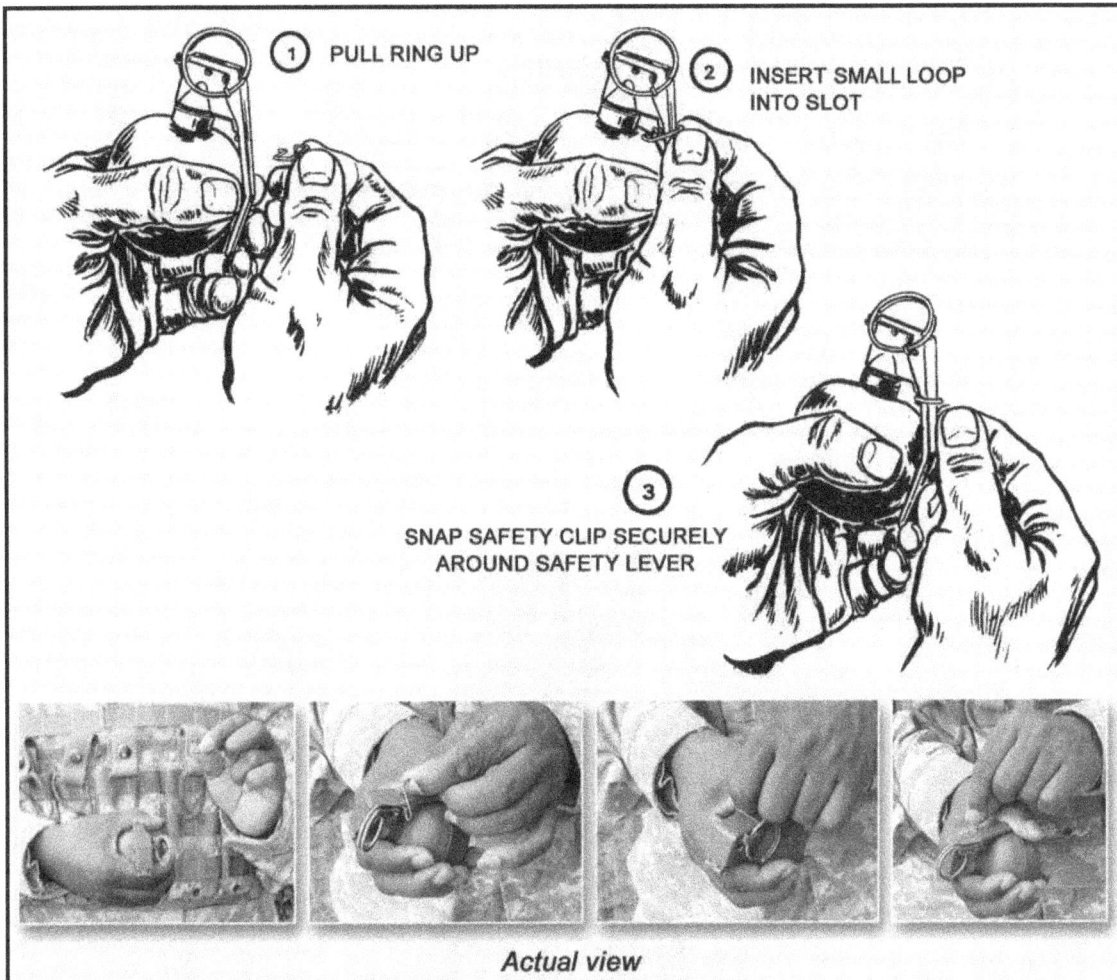

Figure 3-5. Safety clip installation.

DAILY CHECKS

3-11.   Personnel should check hand grenades daily. To perform daily checks (Figure 3-6)—

(1)   Ensure the safety pin (1) is present.

CAUTION

Do not bend the ends of the pull ring safety pin back flush against the fuze body. This practice, intended to preclude the accidental pulling of the pin, makes the removal of the safety pin difficult. Repeated working of the safety pin in this manner causes the pin to break, creating a hazardous condition.

---

(2)   Ensure the confidence clip (2) is present and properly secured to the pull ring.

(3)   Ensure the safety clip (3) is present and properly secured to the safety lever (4).

(4)   Check the grenade fuze assembly (5) for tightness.

```
┌──────────────────────────────────────────────────────────┐
│                        WARNING                            │
│                                                            │
│        Never remove the fuze from a live grenade.          │
│                                                            │
└──────────────────────────────────────────────────────────┘
```

(5) Ensure that the safety lever (4) is not broken.

```
┌──────────────────────────────────────────────────────────┐
│                        WARNING                            │
│                                                            │
│   If the grenade safety lever is broken, do not use the grenade. │
│                                                            │
└──────────────────────────────────────────────────────────┘
```

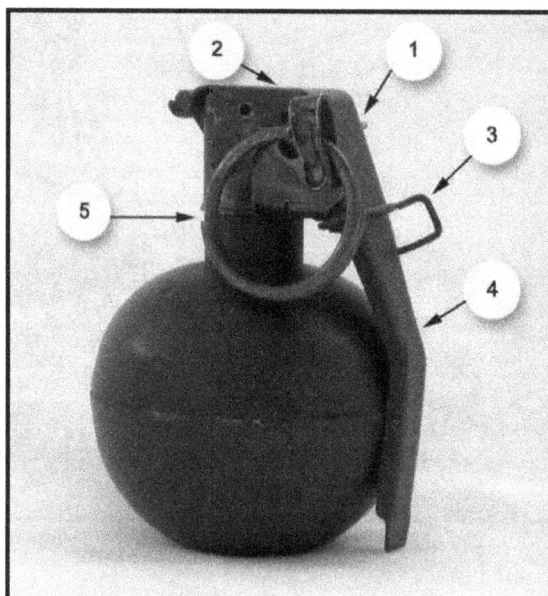

Figure 3-6. Hand grenade safety inspection points.

STORAGE

3-12. Personnel should carry hand grenades on the ammunition pouch using the carrying safety straps designed specifically for this purpose or in the grenade pockets of the ETLBV. When storing grenades in the ammunition pouch or on the ETLBV, personnel should adhere to the following guidelines:

- Ensure that the grenade is fully inside of the carrying pouch and that the pocket flap is fully secured (ETLBV or tactical vest only, shown in Figure 3-7).

**Figure 3-7. Grenade carrying pouch
(attached to enhanced tactical load-bearing vest).**

- Ensure the safety lever is inside of the carrying pouch, and the pull ring is in the downward position. Then, wrap the restraining strap around the neck of the fuze, and secure to the LCE only (Figure 3-8).

Figure 3-8. Grenade carrying pouch (attached to load-carrying equipment).

- DO NOT put adhesive tape around a grenade fuze, the pull ring, or safety lever as is shown below (Figure 3-9).

> **WARNING**
>
> Do not tape a hand grenade safety lever or safety pin. The safety pin can come off with the tape, causing the grenade to explode.

Figure 3-9. Taping and storing hand grenades.

- Carry hand grenades using the proper procedures.

> **WARNING**
>
> Never carry the grenades suspended by the safety pull ring or safety lever.

- DO NOT tape hand grenades to Soldier gear (Figure 3-10).

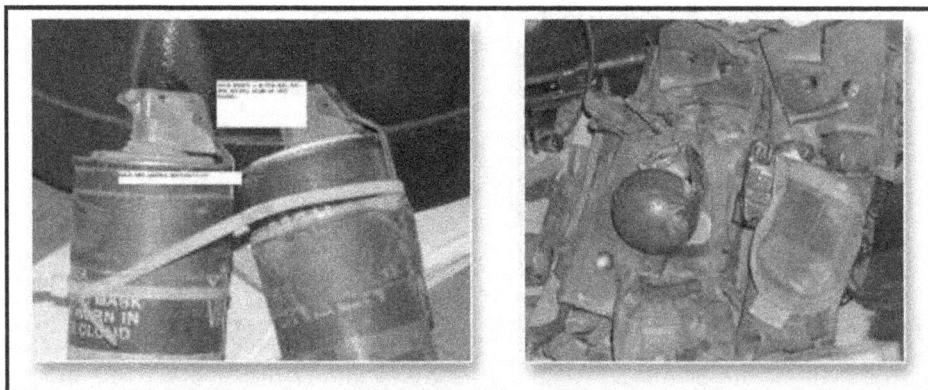

Figure 3-10. Taping hand grenades to Soldier gear.

- Never make unauthorized modifications to hand grenades.

WARNING

Grenade safety devices are designed so that the grenade remains safe in storage. Do not bend, tamper, modify or otherwise alter a hand grenade safety pin or safety lever.

- During air operations in wartime conditions, Soldiers must be prepared to engage the enemy as soon as they land; therefore, Soldiers must carry their grenades in their ammunition pouches.

CAUTION

During training missions, do not carry hand grenades in ammunition pouches during airborne operations. Carry the grenades in the main body of the rucksack instead.

USE

3-13. To safely throw hand grenades, Soldiers must demonstrate and execute the proper techniques of gripping, preparing, and throwing the grenade.

NOTE: The M69 practice hand grenade is used for all individual and collective training tasks.

GRIPPING

3-14. A grenade not held properly is difficult to arm. Gripping procedures differ slightly for right- and left-handed Soldiers; however, proper hold remains constant. Holding the grenade in the throwing hand with the safety lever placed between the first and second joints of the thumb provides safety and throwing efficiency.

Right-Hand Grip

3-15.   To throw the grenade with the right hand, Soldiers should use the right-hand grip (Figure 3-11). To use the right-hand grip, hold the grenade upright, with the pull ring away from the palm of the throwing hand so that the pull ring can be easily removed by the index or middle finger of the free hand.

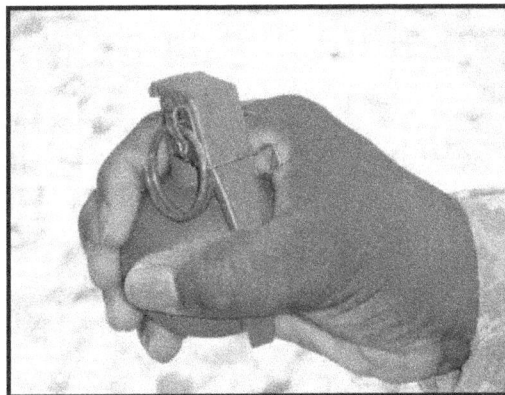

Figure 3-11. Right-hand grip.

Left-Hand Grip

3-16. To throw the grenade with the left hand, Soldiers should use the left-hand grip (Figure 3-12). To use the left-hand grip, invert the grenade, with the pull ring away from the palm of the throwing hand so that the pull ring can be easily removed by the index or middle finger of the free hand.

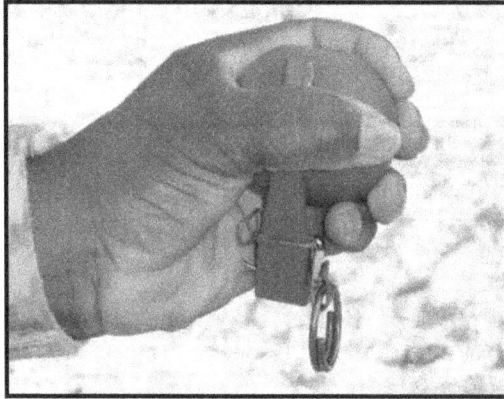

Figure 3-12. Left-hand grip.

PREPARING

NOTE: Do not remove the safety clip or the safety pin until the grenade is about to be thrown.

CAUTION

Never attempt to reinsert a safety pin into a hand grenade during training. In combat, however, it may be necessary to reinsert a safety pin into a grenade. Take special care to replace the pin properly. If the tactical situation allows, it is safer to throw the grenade rather than to trust the reinserted pin.

3-17. The method of grenade preparation is determined by the thrower's grip. To prepare the grenade—
- Tilt the grenade forward to observe the safety clip.
- Remove the safety clip by sweeping it away from the grenade with the thumb of the opposite hand (Figure 3-13 and Figure 3-14).

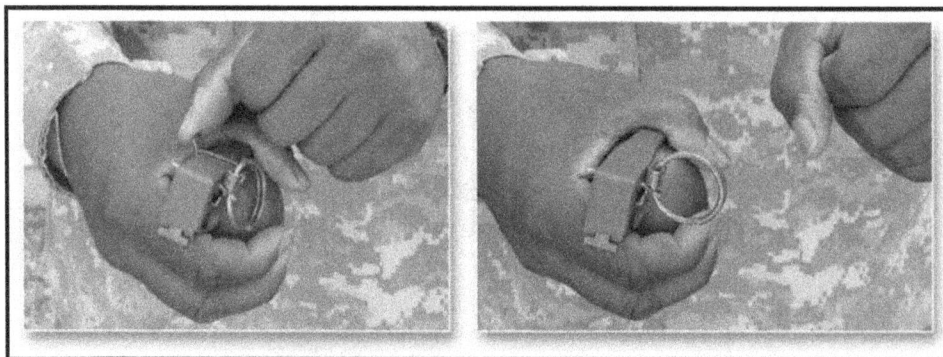

Figure 3-13. Right-hand grip, removing the safety clip.

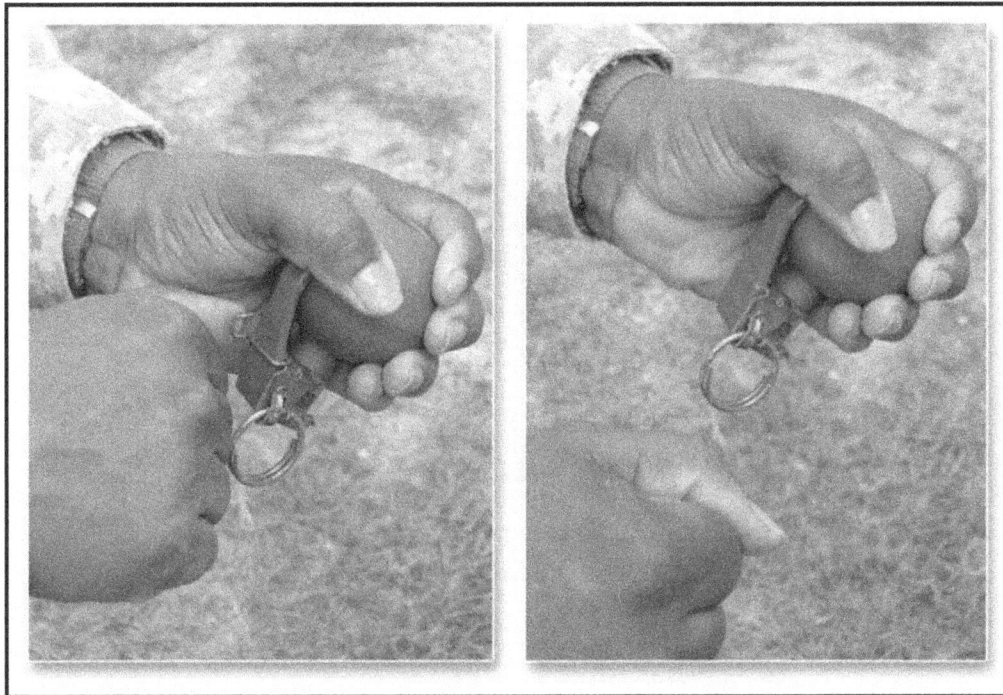

Figure 3-14. Left-hand grip, removing the safety clip.

- Insert the index or middle finger of the nonthrowing hand in the pull ring until it reaches the knuckle of the finger (Figure 3-15).

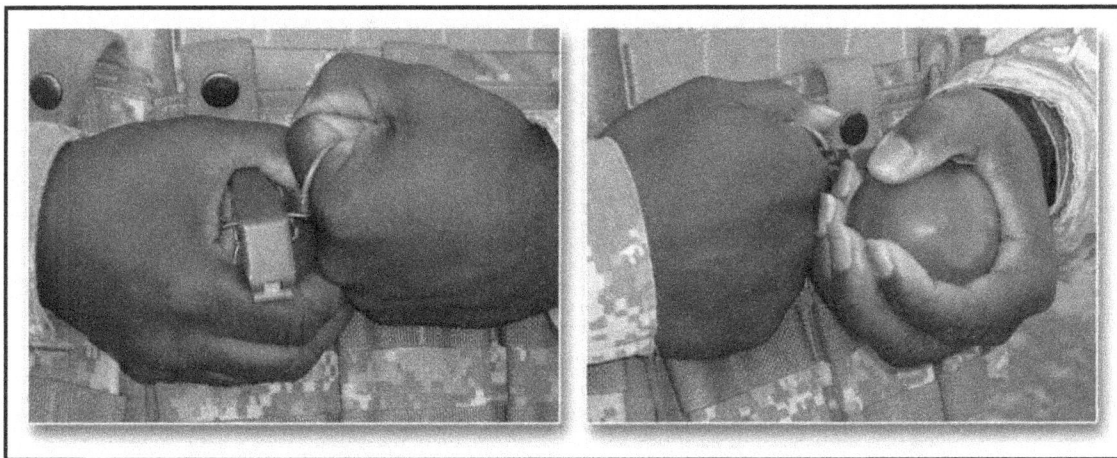

Figure 3-15. Pull ring grip, right-/left-hand thrower.

(1) Remove the safety pin from the confidence clip. For a right-handed thrower (Figure 3-16)—

- Twist the pull ring toward the body to release the pull ring from the confidence clip.
- Pull the safety pin from the grenade.

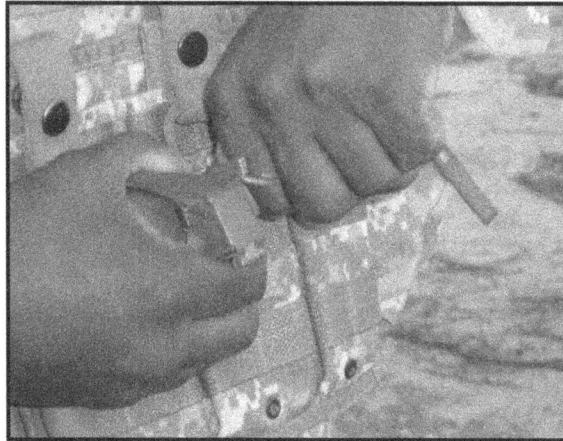

Figure 3-16. Right-hand grip, pulling the safety pin.

For a left-handed thrower (Figure 3-17)—

- Twist the pull ring away the body to release the pull ring from the confidence clip.
- Pull the safety pin from the grenade.

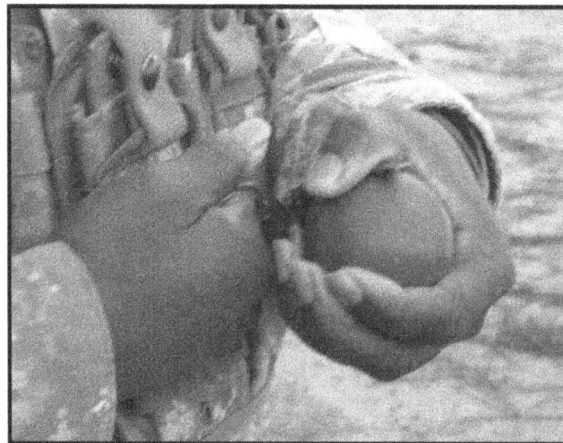

Figure 3-17. Left-hand grip, pulling the safety pin.

(2) Ensure that you are holding the safety lever down firmly.

> **DANGER**
> IF PRESSURE ON THE SAFETY LEVER IS RELAXED AFTER THE SAFETY CLIP AND SAFETY PIN ARE REMOVED, THE STRIKER CAN ROTATE AND STRIKE THE PRIMER WHILE THE THROWER IS STILL HOLDING THE GRENADE. CONTINUING TO HOLD THE GRENADE BEYOND THIS POINT CAN RESULT IN INJURY OR DEATH.

(3) Remove the safety pin by pulling on the pull ring.

COOKING OFF

> ### WARNINGS
>
> **In training, never cook off live fragmentation hand grenades or offensive concussion grenades. Use cook-off procedures only in a combat environment.**
>
> **Never cook off the M84, stun grenade, or smoke grenades. These grenades have short fuze delays (1 to 2.3 seconds) and will cause serious personal injury if cook-off procedures are performed.**

3-18. To achieve above-ground detonation or near-impact detonation—

   (1) Remove the grenade's safety pin.

   (2) Release the safety lever.

   (3) Count "One thousand one, one thousand two."

   (4) Throw the grenade.

3-19. This is called cooking off. Cooking off uses enough of the grenade's 4- to 5-second delay (about 2 seconds) to cause the grenade to detonate above ground or shortly after impact with the target.

THROWING

> ### WARNING
>
> **Throwers must consider the flight path of the grenade to make sure no obstacles alter the flight of the grenade or cause it to bounce back toward them.**

3-20. Soldiers can use five positions to throw grenades:

- Standing.
- Prone-to-standing.
- Kneeling.
- Prone-to-kneeling.
- Alternate prone.

3-21. Tactical employment of the hand grenade is METT-TC dependant. The situation will usually dictate the position best suited for delivering a grenade on target.

3-22. If a Soldier can achieve more distance and accuracy using his own personal style, he should be allowed to do so as long as his body is facing sideways and toward the enemy's position, and he throws the grenade overhand. There are, however, general steps that the Soldier must follow:

   (1) Observe the target to estimate the distance between the throwing position and the target area.

NOTE: In observing the target, minimize exposure time to the enemy (no more than 3 seconds).

   (2) Grip the hand grenade in the right or left throwing hand.

   (3) Look at the target.

   (4) Throw the grenade overhand so that the grenade arcs, landing on or near the target.

(5) Allow the motion of the throwing arm to continue naturally once the grenade is released.

NOTE: This follow-through improves distance and accuracy and lessens the strain on the throwing arm.

---

(6) Seek cover to avoid being hit by fragments or direct enemy fire. If no cover is available, drop to the prone position with your protective head gear facing the direction of the grenade's detonation.

NOTE: Soldiers should practice throws that are used in combat, such as the underhand and sidearm throws. Soldiers can practice these throws with practice grenades, but they must throw live fragmentation grenades overhand in a training environment.

Standing Position

3-23.   The standing position is the most desirable and natural position from which to throw grenades. It allows the Soldier to obtain the greatest possible throwing distance. However, this position should only be used when cover and concealment is readily available.

3-24.   To throw a grenade from the standing position (Figure 3-18)—

(1) Observe the target to estimate the distance between the throwing position and the target area.

(2) Assume a natural stance, with your weight balanced equally on both feet.

(3) Prepare the grenade.

(4) Hold the grenade shoulder high and the nonthrowing hand at a 45-degree angle with the fingers and thumb extended, joined, and pointing toward the intended target.

(5) Throw the grenade overhand so that the grenade arcs, landing on or near the target.

(6) Allow the motion of the throwing arm to continue naturally once the grenade is released.

(7) Seek cover to avoid being hit by fragments or direct enemy fire. If no cover is available, drop to the prone position facing the direction of the grenade's detonation.

Figure 3-18. Standing position.

Prone-To-Standing Position

3-25.   When exposure time is more important than accuracy and cover and concealment is not readily available, the prone-to-standing position can be used to immediately suppress an area.

3-26.   To throw a grenade from the prone-to-standing position (Figure 3-19)—

(1) Lie down on the stomach with the body parallel to the grenade's intended line of flight.

(2) Hold the grenade at chest level.

(3) Place the hands in a push-up position, and stand up while holding the grenade in the throwing hand. Assume a good standing position, if the situation permits.

(4) Prepare the grenade.

(5) Hold the grenade shoulder high and the nonthrowing hand at a 45-degree angle with the fingers and thumb extended, joined, and pointing toward the intended target.

(6) Throw the grenade overhand so that the grenade arcs, landing on or near the target.

(7) Allow the motion of the throwing arm to continue naturally once the grenade is released.

(8) After throwing the grenade, drop to the ground on the stomach and press flat against the ground.

Figure 3-19. Prone-to-standing position.

Kneeling Position

3-27.   The kneeling position reduces the distance a Soldier can throw a grenade. It is used primarily from behind low-level ground cover.

3-28.   To throw a grenade from the kneeling position (Figure 3-20)—

(1) Observe the target to mentally estimate the throwing distance.

(2) Prepare the grenade while behind cover.

(3) Bend the nonthrowing knee at a 90-degree angle, placing that knee on the ground. Keep the throwing leg straight and locked, with the side of the boot firmly on the ground.

(4) Move the body to face sideways, toward the target position.

(5) Hold the grenade shoulder high and the nonthrowing hand at a 45-degree angle with the fingers and thumb extended, joined, and pointing toward the intended target.

(6) Throw the grenade overhand so that the grenade arcs, landing on or near the target. Push off with the throwing foot to give added force to the throw.

(7) Allow the motion of the throwing arm to continue naturally once the grenade is released.

(8) Drop to the prone position or behind available cover to reduce exposure to fragmentation and direct enemy fire.

Figure 3-20. Kneeling position.

Prone-To-Kneeling Position

3-29.   The prone-to-kneeling position enables the Soldier to throw the grenade farther and is performed for the same reason as the prone-to-standing position; time to throw is more important than accuracy.

3-30.   To throw a grenade from the prone-to-kneeling position (Figure 3-21)—

    (1) Lie down on the stomach, with the body parallel to the grenade's intended line of flight.

    (2) Hold the grenade at chest level.

    (3) Place the hands in a push-up position, and assume the kneeling position while holding the grenade in the throwing hand. Assume a good kneeling position, if the situation permits.

    (4) Prepare the grenade.

    (5) Hold the grenade shoulder high and the nonthrowing hand at a 45-degree angle with the fingers and thumb extended, joined, and pointing toward the intended target.

    (6) Throw the grenade overhand so that the grenade arcs, landing on or near the target.

    (7) Allow the motion of the throwing arm to continue naturally once the grenade is released.

    (8) After throwing the grenade, drop to the ground on the stomach and press flat against the ground.

Figure 3-21. Prone-to-kneeling position.

Alternate Prone Position

3-31. The alternate prone position reduces both distance and accuracy and is used when rising to engage a target is not safe.

3-32. To throw a grenade from the alternate prone position (Figure 3-22)—

   (1) Lie down on the back, with the body parallel to the grenade's intended line of flight.
   (2) Hold the grenade at chin/chest level.
   (3) Prepare the grenade.
   (4) Cock the throwing leg at a 45-degree angle, maintaining knee-to-knee contact and bracing the side of the boot firmly on the ground.
   (5) Hold the grenade 4 to 6 inches behind the ear with the arm cocked for throwing.
   (6) With the free hand, grasp any object that will provide additional leverage to increase the throwing distance.
   (7) Throw the grenade, and push off with the rearward foot to give added force to the throw.

CAUTION

Do not lift the head or body when attempting to throw the grenade as this may cause exposure to direct enemy fire.

---

(8) After throwing the grenade, roll over onto the stomach and press flat against the ground.

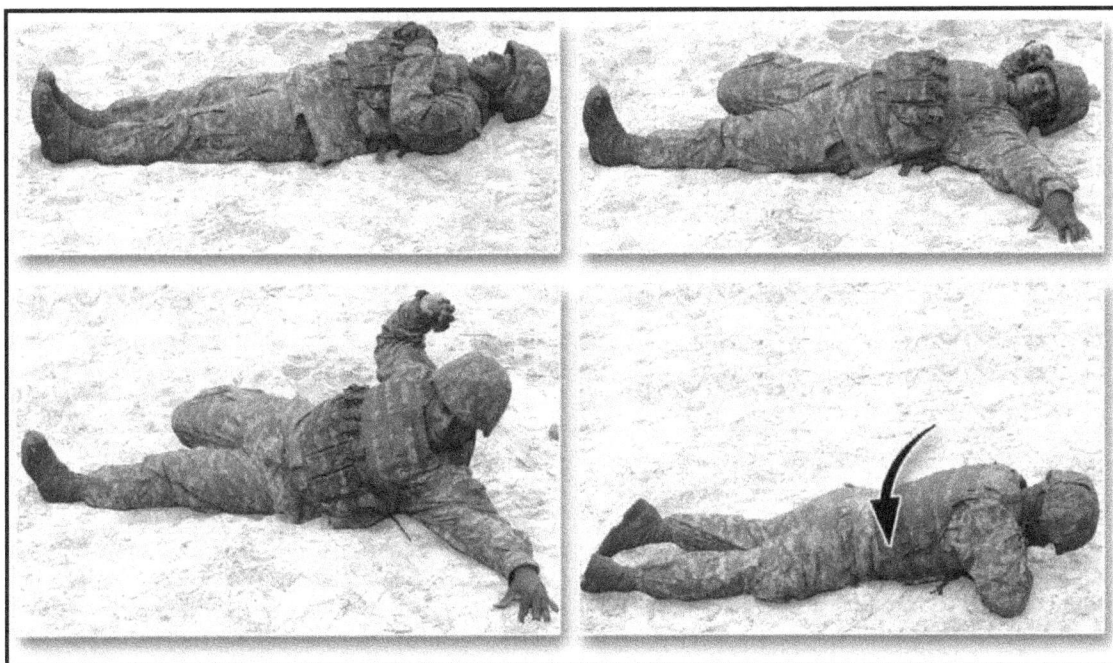

Figure 3-22. Alternate prone position.

MAINTENANCE

3-33. When exposed to the environment, hand grenades require just as many preventive maintenance checks and services (PMCS) as a Soldier's personal weapon. The body of the hand grenade is made of metal, which rusts when it is exposed to moisture or submerged in water. If not removed, dirt or rust can cause the hand grenade to malfunction.

NOTE: See TM 9-1330-200-12 for more information about required grenade maintenance.

3-34. For most hand grenades, keeping them clean and lubricated is sufficient maintenance. With the M69 practice grenade, however, maintenance is more difficult because the grenade bodies are used repeatedly. To maintain the M69 practice grenade—

- Paint the grenade body at least quarterly.
- Clean the threads with a wire brush on a monthly basis.
- Remove fuze residue from the body immediately after each use.

NOTE: Cleaning the threads and removing the residue from the hand grenade body make replacement of the fuzes easier. The grenade body lasts longer if these preventive maintenance procedures are performed.

CLEANING

3-35. To clean the grenade—

- Wipe the dirt off the body of the hand grenade using a slightly damp cloth or a light brush.
- Use a light brush to clean the fuze head, as it can reach into the crevices.

CAUTION

Vigorous cleaning of a grenade with a heavy bristled brush or cleaning rag may loosen or dislodge the pull ring.

LUBRICATING

3-36. Depending on weather conditions, a light coat of CLP may be needed.

DESTRUCTION PROCEDURES

3-37. Destruction of any military weapon is authorized only as a last resort to prevent the enemy from capturing or using it. In combat, the commander has the authority to destroy weapons, but he must report doing so through the proper channels.

NOTE: See TM 9-1330-200-12 for more information.

3-38. The conditions under which destruction will be effected are command decisions and may vary depending upon a number of factors, such as—

- Tactical situation.
- Security classification.
- Quantity and location of grenades.
- Facilities for accomplishing destruction.
- Time.

METHODS OF DESTRUCTION

3-39. Selection of the method of destruction requires imagination and resourcefulness in the utilization of the facilities at hand under the existing conditions. In general, destruction of grenades can be accomplished most effectively by burning or detonation, or a combination of these methods.

NOTE: For the successful execution of methods of destruction involving the use of demolition materials, all personnel concerned must be thoroughly familiar with the provision of FM 3-34.214. Training and careful planning are essential.

3-40. If destruction of grenades is directed, due consideration should be given to the following:
  (1) Selection of a site (for the destruction operation) that will cause greatest obstruction to enemy movement and also prevent hazard to friendly troops from fragments incidental to the destruction.
  (2) Observance of appropriate safety precautions.

Detonating

3-41. Packed and unpacked high-explosive (HE) grenades, fuzes, and accessories may be destroyed by placing them in piles and detonating them with demolition charges.

3-42. To use this method—
    (1) Prepare the demolition charge (using one-pound TNT blocks [or equivalent] and the necessary detonating cord to make up each charge).

NOTE: One hundred pounds of packed HE grenades require a two-pound demolition charge to ensure complete detonation of the pile. For unpacked HE grenades, a one-pound demolition charge is sufficient.

---

    (2) Place the charges on the pile to be detonated.
    (3) Provide for dual priming to minimize the possibility of a misfire. For priming, use a non-electric blasting cap crimped to at least 5 feet of time blasting fuze or an electric blasting cap and firing wire).

NOTE: Time blasting fuzes contain black powder and a blasting cap. They must be protected from moisture at all times.

WARNING

Each roll of fuze must be tested shortly before use. The burning rate of a safety fuze varies under different atmospheric and/or climatic conditions (from a burning time of 30 seconds or less per foot to 45 seconds or more per foot).

WARNING

Blasting caps, detonating cord, and time blasting fuzes must be kept separated from the charges until required for use.

    (4) If primed with non-electric blasting cap and time blasting fuze, ignite and take cover; if primed with electric blasting cap, take cover before firing the charge.

NOTE: Time blasting fuzes may be ignited by a blasting fuze igniter or an ordinary match; the electric blasting cap requires a blasting machine or equivalent source of electricity.

3-43. The danger area for piles detonated in the open is a circular area which varies according to the quantity of explosive items to be destroyed.

NOTE: Quantity/distance data (inhabited building distance) is given in Chapter 4 of TM 9-1330-200-12.

Burning

3-44. Packed and unpacked HE grenades, smoke grenades, and illuminating grenades may be destroyed quickly and effectively by burning.

3-45. To destroy hand grenades by fire—
    (1) Stack the ammunition into a pile.
    (2) Place flammable materials (e.g., rags, scrap wood, or brush) on and about the pile.
    (3) Pour gasoline and oil over the entire pile.

WARNING

Consideration should be given to the highly flammable nature of gasoline and its vapor. Carelessness in its use may result in painful burns.

(4) Ignite the pile using an incendiary grenade fired from a safe distance, a combustible train of suitable length, or other appropriate means.

(5) Take cover immediately.

WARNING

Cover must be taken without delay, since an early explosion of ammunition may be caused by the fire.

3-46. The danger area for piles being burned in the open is 600 meters.

DEGREE OF DAMAGE

3-47. The method of destruction used must damage the grenades and their components to such an extent that they cannot be restored to usable condition in the combat zone. Further, the same essential components of all grenades must be destroyed so that the enemy cannot assemble complete rounds from undamaged components of several damaged complete rounds.

Chapter 4
Pyrotechnic Signals and Simulators

Pyrotechnic signals can be used as a means of communication and signals, obscuration, warning of an intruder, and for simulating enemy fires.

SECTION I. COMMUNICATION SIGNALS

There are two classifications of pyrotechnic communication signals: handheld signals and ground smoke signals. Both types of signals come in varied color patterns. Soldiers can use these patterns to coordinate troop movements and, in the case of an emergency, designate pick-up points.

NOTE: The signals are usually prescribed at command level and prearranged in accordance with signal operating/operation instructions (SOI).

HANDHELD SIGNALS

4-1.   Star clusters, star parachutes, and smoke parachutes are issued in an expendable launcher that consists of a launching tube and firing cap (Figure 4-1).

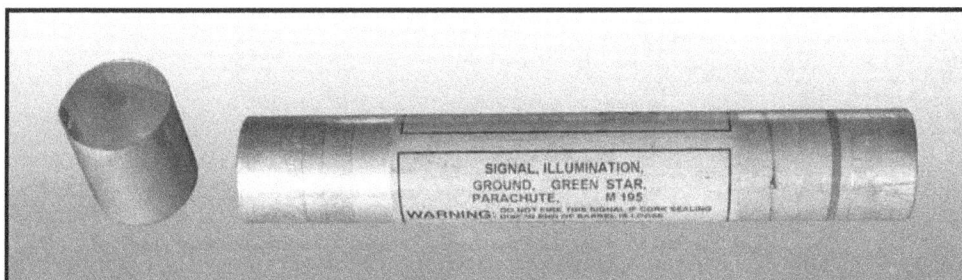

Figure 4-1. Handheld pyrotechnic signal.

INSPECTION

WARNING

Handheld communication signals must be inspected before use and properly secured to avoid serious injury.

NOTE: See TM 9-1370-206-10 for more information about handheld communication signal inspection.

Initial Inspection

> **CAUTION**
>
> During storage, keep boxes sealed. Duds or improper functioning could occur if exposed to moisture for long periods of time. Open just before use.

4-2.   When in bulk, communication signals are secured in shipping containers (Figure 4-2). Personnel should inspect the shipping container upon receipt. Shipping containers that are damaged should not be opened; they should be returned to the ASP or disposed of using the methods outlined in the unit SOP.

4-3.   There are two types of communication signal shipping containers: the M548 metal container and wood ammunition box containers.

M548 metal container

4-4.   This container contains 24-handheld signals, individually secured in plastic containers (Figure 4-2).

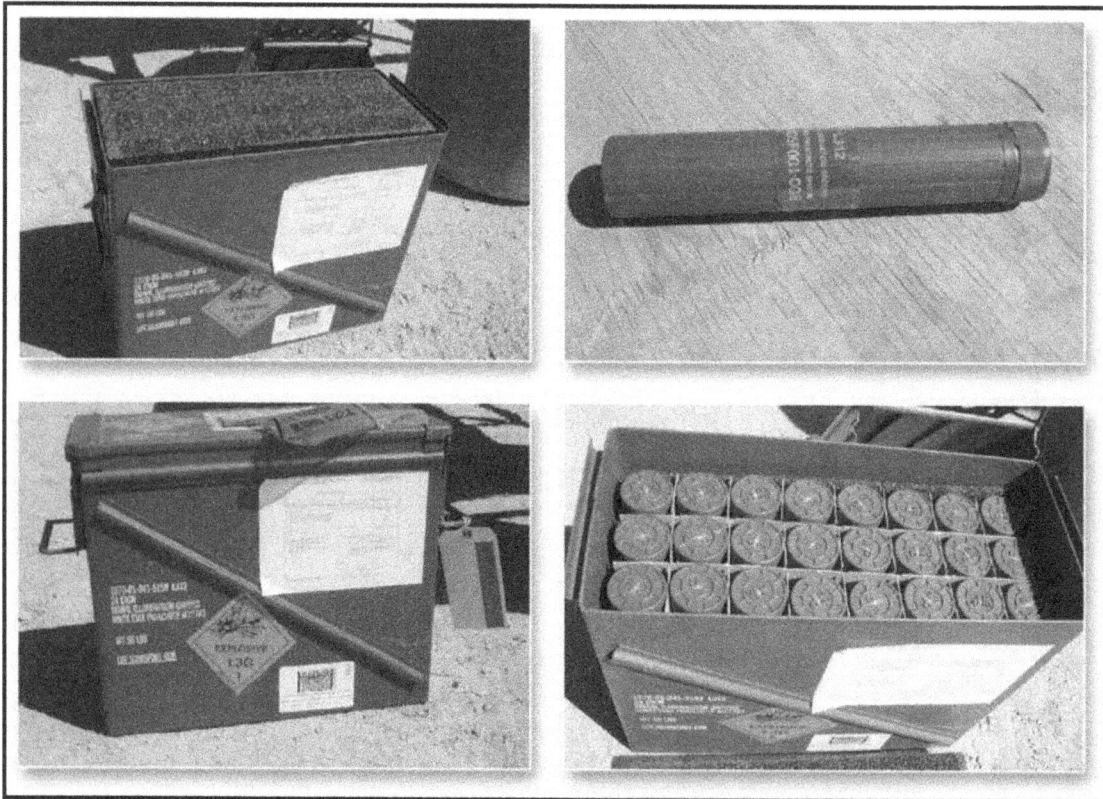

Figure 4-2. Handheld signal shipping container.

Wood AmmunitionBox Container

4-5.   This container contains 36-handheld signals, sealed in plastic barrier bags (Figure 4-3). Each barrier bag contains 18-hermetically sealed metal containers. Each sealed container contains one handheld signal.

4-6.   Upon removing the sealed barrier bags from the shipping container (Figure 4-3), personnel should inspect each barrier bag and identify any of the following discrepancies:

- The barrier bag has been damaged.
- The seal on the barrier bag show signs of tampering.

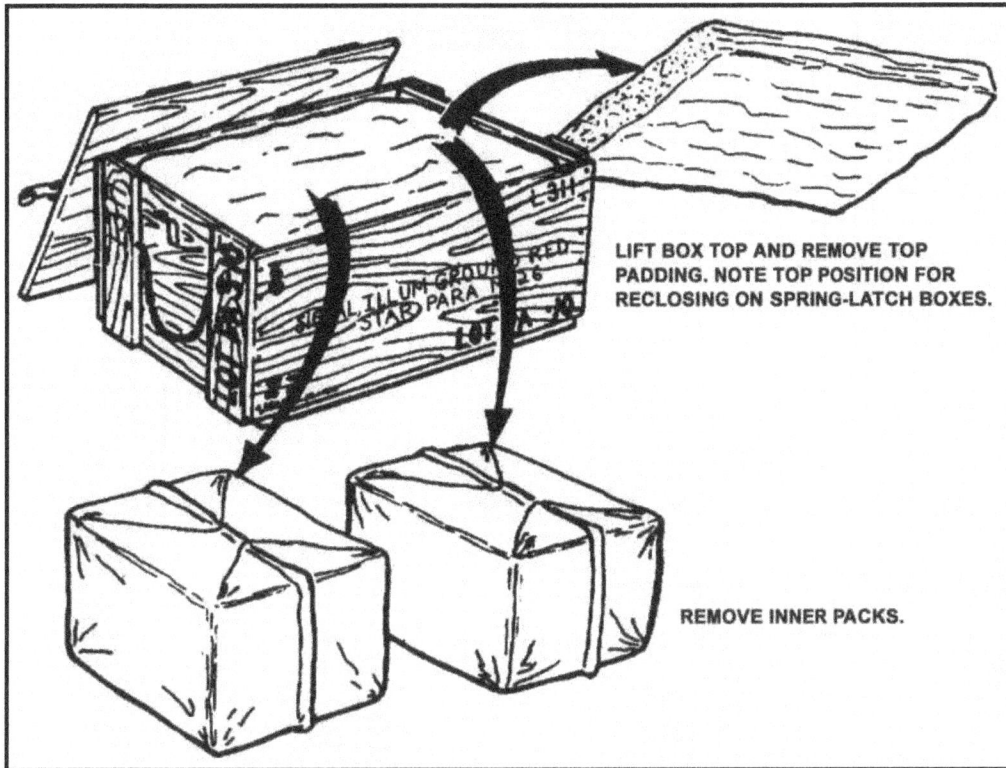

LIFT BOX TOP AND REMOVE TOP PADDING. NOTE TOP POSITION FOR RECLOSING ON SPRING-LATCH BOXES.

REMOVE INNER PACKS.

Figure 4-3. Handheld signal barrier bag.

4-7.   Each signal within the barrier bag is housed in a hermetically sealed steel container (Figure 4-4) or plastic container (Figure 4-5). Handheld communication signals may be issued while still in their individual containers or unpacked and issued by someone within the chain of command. Upon removing the sealed individual containers from the barrier bag (Figure 4-4), personnel should inspect the containers and identify any of the following discrepancies:

- The container has been damaged.
- The seal on the container has been tampered with or is missing.

CAUTION

Do not open hermetically sealed (air tight) containers until ready for use. A signal exposed to moisture may not function.

4-8.   Then, personnel should remove the handheld signal from the container.

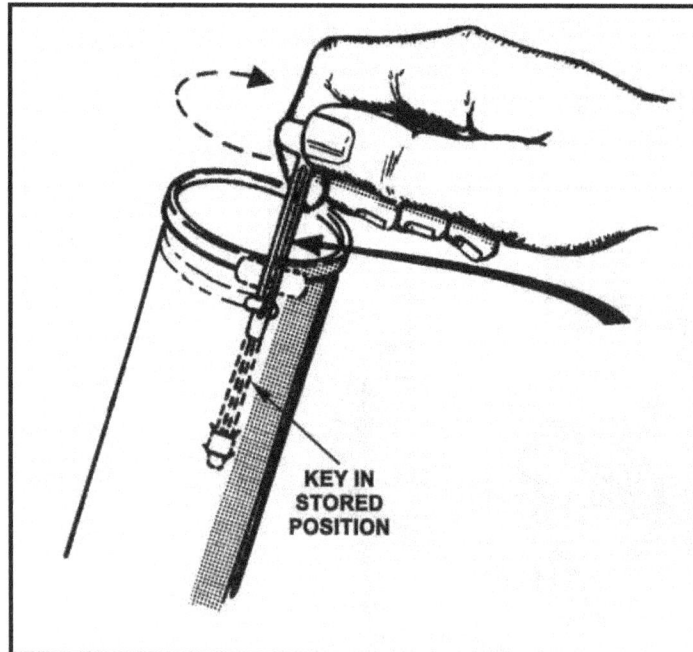

Figure 4-4. Handheld signal individual sealed steel container.

Handheld Signal Sealed Steel Container

4-9.   To open the handheld signal from a hermetically sealed steel container, by using the key attached to the container (Figure 4-4)—

    (1) Remove the sealing strip.

    (2) Remove the top of container.

    (3) Remove any padding pieces from the container.

    (4) Remove the signal.

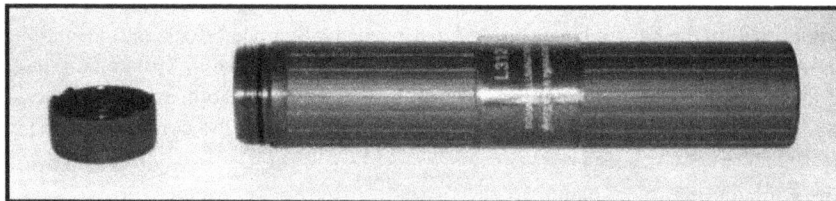

Figure 4-5. Handheld signal individual container.

Handheld Signal Plastic Sealed Container

4-10. To open the handheld signal plastic sealed container (Figure 4-5)—

    (1) Hold the container in one hand.

    (2) Twist the end-cap counterclockwise with the other hand.

    (3) Remove the signal.

Inspect a Handheld Signal

4-11. Once the signal has been removed from the container (Figure 4-6), personnel should inspect the signal and identify any of the following discrepancies:

    ● Corrosion is on the launcher tube.

    ● Holes are visible in the launcher tube.

- The forward-end seal is broken or damaged.
- The firing pin is not present.
- The primer is not intact (is dented).
- The color-coded forward-end seal does not match the color listed on the data plate.

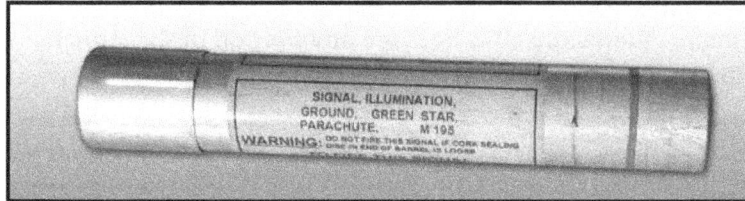

Figure 4-6. Handheld signal removed from the individual container.

NOTE: If any of the discrepancies are found upon receipt of newly issued handheld signals, personnel should return the signal and individual container to the issuing person or dispose of it in accordance with the unit SOP.

Before Storing

4-12. Before securing handheld communication signals, personnel should take the following safety precautions (Figure 4-7):
- Ensure the launcher tube is not bent or punctured.
- Check the launcher tube for corrosion or dirt.
- Ensure that the forward-end seal is not broken or damaged.
- Ensure that the firing pin is present and the primer is intact (not dented).

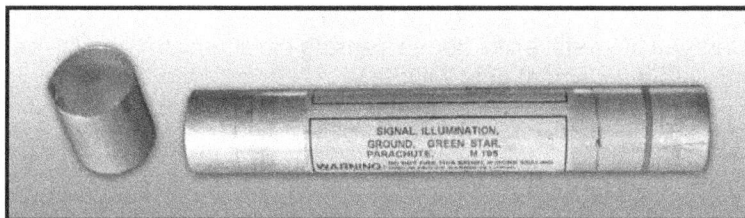

Figure 4-7. Safety inspection points—before storage.

Daily Checks

4-13. Personnel should check handheld communication signals daily to ensure that they are free of foreign material and that they are not damaged.

STORAGE

4-14. Personnel should carry handheld communication signals in accordance with the unit SOP. When carrying handheld communication signals, personnel should adhere to the following guidelines:
- Ensure that the communication signal is placed in a secure, dry area.

CAUTION

Duds or improper functioning could occur if exposed to moisture for long periods of time. Open just before use.

● Do not put adhesive tape around any portion of the handheld communication signal.

WARNING

Do not bend, tamper, modify or otherwise alter a handheld communication signal. Do not tape any portion of the launcher or firing mechanism.

USE

4-15. To safely use handheld signals, Soldiers must properly determine the type and color of the pyrotechnic signal to be used, and safely launch the signal.

Determining the Type and Color

4-16. When choosing a pyrotechnic signal, Soldiers must consider the signal's intensity and color.

Intensity

4-17. Handheld colored pyrotechnic signal flares burn at different intensities.

Color

4-18. Determining specific colors at night is not difficult. However, Soldiers should avoid using red and green star clusters near aircraft.

CAUTION

Avoid signaling aircraft at night with star clusters. Red and green star clusters can be mistaken for tracers causing the aircraft to open fire on the friendly ground element or to withdraw.

4-19. In daylight, Soldiers should adhere to the following considerations:
● It can be difficult to differentiate between white and green depending on lighting conditions.
   ■ Green is very pale in daylight and is especially difficult to detect in fog, haze, or smoke-filled skies. In fact, white flares are easier to detect in daylight than green.
   ■ White flares can be mistaken for illumination flares.
● Red may be difficult to detect when launched in a position that forces the observer to see it near a vivid sunrise or sunset.

Launching Handheld Signals

NOTE: See TM 9-1370-206-10 for more detailed information on safety precautions.

4-20. To fire handheld signals (Figure 4-8)—
(1) Observe the surrounding area to ensure that you have overhead clearance.

DANGER

DO NOT FIRE A HANDHELD SIGNAL IN AN AREA WITHOUT OVERHEAD CLEARANCE. WHEN FIRED IN AN AREA WITHOUT OVERHEAD CLEARANCE, THE SIGNAL CAN CAUSE FIRE, INJURY, OR DEATH.

(2) Grasp the signal firmly with your nonfiring hand, red-knurled band down, with your little finger above the red band.

(3) With your firing hand, withdraw the firing cap from the upper end of the signal.

(4) Point the ejection end of the signal up and away from your body, and push the firing cap onto the signal until the open end of the cap is aligned with the red band.

(5) Hold the signal away from your body and at the desired trajectory angle.

WARNING

Turn your head away from the signal to avoid injury to your face and eyes from particles ejected by the small rockets.

Figure 4-8. Firing a handheld signal.

(6) Strike the bottom of the cap using a sharp blow with the palm of your firing hand or strike it on a hard surface, keeping your nonfiring arm rigid.

CAUTION

When firing handheld signals by hand, avoid contact with the bones of the hand. This can result in injury to the hand. Instead, use the meaty portion of the hand.

Figure 4-8. Firing a handheld signal (continued).

Misfire

4-21. In the event of a misfire—

(1) While keeping the signal aimed, pull the cap back to the red knurled band, and rotate 90 degrees.

(2) Make two more attempts to fire.

(3) If it still does not fire, wait 30 seconds keeping the arm rigid and the signal aimed overhead.

(4) Return the cap to the ejection end of the signal and dispose of it in accordance with unit SOP.

MAINTENANCE

4-22. When exposed to the environment, handheld signals require PMCS. The color-coded forward end seal can deteriorate if exposed to moisture for long periods of time or submerged in water. If not removed, dirt or sand can cause the handheld signal to malfunction.

NOTE: See TM 9-1370-206-10 for more information about required maintenance.

Cleaning

4-23. To clean the handheld signal—

● Wipe the dirt off the launcher tube and the firing cap using a clean, dry, lint-free cloth.

● Use a fine-bristled camel hair brush to remove any foreign matter or debris.

DESTRUCTION PROCEDURES

4-24. Destruction of any military weapon is authorized only as a last resort to prevent the enemy from capturing or using it. In combat, the commander has the authority to destroy weapons, but he must report doing so through the proper channels.

4-25. The conditions under which destruction will be effected are command decisions and may vary depending upon a number of factors, such as—

● Tactical situation.

● Security classification.

● Quantity and location of grenades.

● Facilities for accomplishing destruction.

● Time.

METHODS OF DESTRUCTION

4-26. Selection of the method of destruction requires imagination and resourcefulness in the utilization of the facilities at hand under the existing conditions. In general, destruction of handheld signals can be accomplished most effectively by burning or firing, or a combination of these methods.

DEGREE OF DAMAGE

4-27. The method of destruction used must damage the handheld signals and their components to such an extent that they cannot be restored to usable condition in the combat zone. Further, the same essential components of all handheld signals must be destroyed so that the enemy cannot assemble complete rounds from undamaged components of several damaged complete signals.

GROUND SMOKE SIGNALS

4-28. Ground smoke signals, or smoke grenades, come in various colors (Figure 4-9). They can be used as a ground-to-ground or ground-to-air signaling device, to convey information through a prearranged signal, or for screening unit movements. These smoke grenades have a 0.7- to 2.0-second time delay and produce a smoke cloud that lasts approximately 13 to 30 seconds.

NOTE: See TM 9-1330-200-12 for more information about smoke grenades.

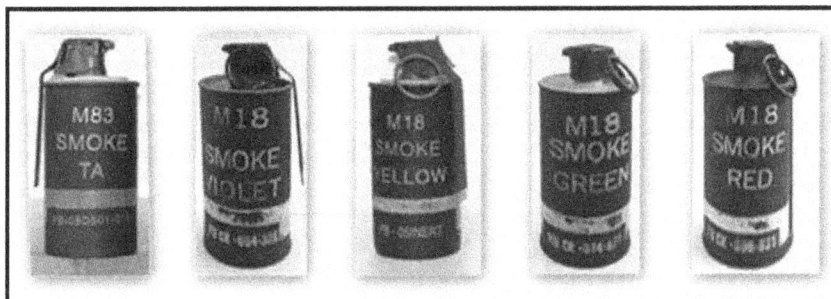

Figure 4-9. Smoke grenades.

INSPECTION

WARNING

Smoke grenades must be inspected before use and properly secured to avoid serious injury.

NOTE: See Chapter 3 for inspection procedures.

STORAGE

4-29. Personnel should secure smoke grenades using a similar method to the carrying of hand grenades.

NOTE: See Chapter 3 for storage procedures.

USE

4-30. Before employing smoke grenades, Soldiers must understand the effects of environmental conditions on obscuration.

CONSIDERATIONS

4-31. Soldiers should consider wind direction and speed before employing smoke grenades:

- Throw grenades upwind of the desired location.
- Lack of wind and heavy humidity can cause smoke to linger. Heavy concentrations of smoke can obscure obstacles, restrict friendly movements, and hide an enemy's location.
- Do not throw smoke grenades on dry tender.

CAUTION

M18, M83, and AN-M8 HC have the potential to start fires when thrown on dry tender.

EMPLOYMENT

4-32. To safely throw smoke grenades, Soldiers must demonstrate and execute the proper techniques of gripping, preparing, and throwing the grenade.

NOTE: Smoke grenades have a pull ring, safety pin, and safety lever. At some time, the confidence clip will be added to all smoke grenades. Refer to Chapter 3 for the technique used to remove the pull ring secured by a confidence clip.

Gripping

4-33. Gripping procedures for the smoke grenade are the same as those for the M67 fragmentation grenade.

NOTE: See Chapter 3 for gripping procedures.

Preparing

4-34. Preparing procedures for the smoke grenade are the same as those for the M67 fragmentation grenade.

NOTE: See Chapter 3 for preparing procedures.

Throwing

4-35. Many of the same throwing positions that are used to employ hand grenades can also be used to employ smoke.

NOTE: See Chapter 3 for more information about hand grenade throwing positions.

SECTION II. TRIP FLARES

The M49 trip flare resembles a hand grenade in size and shape, except that it is provided with a bracket for attachment to a tree or post and a trigger mechanism for firing. Surface trip flares (Figure 4-10) can be used to—

- Provide early warning of infiltration of enemy troops or signaling.

NOTE: To use the surface trip flare as an early warning device, Soldiers should attach a trip wire to the trigger or pull pin. This arms the flare.

- Illuminate an immediate area.

NOTE: Trip flares are not suitable for producing continuous illumination.

- Ignite fires.
- Identify firing ports.
- Force the enemy to withdraw.
- Destroy small, sensitive pieces of equipment (in the same manner as an incendiary grenade).

Figure 4-10. M49A1 surface trip flare.

WARNING

Trip flares must be inspected before use and properly secured to avoid serious injury.

4-36. Several portions of the trip flare must be inspected:
- Body.
- Mounting bracket.
- Other components.

BODY

4-37. The body of the trip flare should be inspected in the same way as that of a hand grenade.

NOTE: See Chapter 3 for inspection procedures.

MOUNTING BRACKET

4-38. Personnel should inspect the mounting bracket to ensure that—

- The mounting bracket is present and firmly affixed to the grenade.
- The mounting bracket shows no signs of damage.

4-39. To inspect the flare and mounting bracket for proper operation (Figure 4-11), press lever against the flare body and check for:

- Straightness of the pull pin.
- Alignment of the safety clip and holes in the cover loading assembly.
- Corrosion and looseness of the cover loading assembly.
- Alignment of the hinge pins in the cover loading assembly.

WARNING

Turn in immediately to the supervisor if the cover assembly is corroded or loose, or if the hinge pins are misaligned.

- Deformed flare and bracket.
- Tension and position of the trigger spring. The trigger should rotate counterclockwise to the extreme position as shown in Figure 4-11, and return to the original position when released.
- Legibility of ammunition lot number.

NOTE: Flares unable to pass the above inspection should be disposed of in accordance with unit SOP.

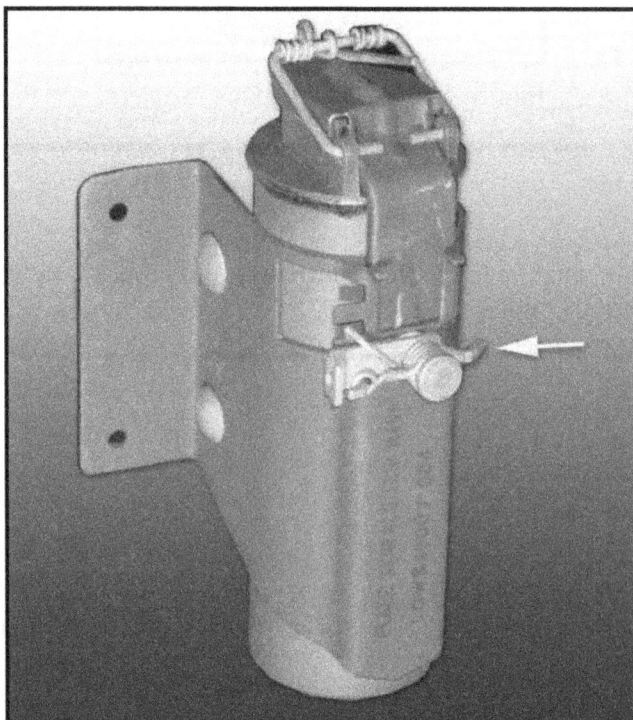

Figure 4-11. M49A1 surface trip flare trigger spring position.

OTHER COMPONENTS

4-40. The trip flare is issued with a spool of trip wire and nails. Personnel should inspect these components and ensure that they are present.

STORAGE

4-41. Personnel should carry trip flares in accordance with the unit SOP. When carrying trip flares, personnel should adhere to the following guidelines:

- Ensure that the trip flare is placed in a secure, dry area.

CAUTION

Duds or improper functioning could occur if exposed to moisture for long periods of time. Open just before use.

- Do not put adhesive tape around any portion of the trip flare during storage.

WARNING

Do not bend, tamper, modify or otherwise alter a trip flare. Do not tape any portion of the trip flare during storage.

USE

NOTE: There is a graphic training aid (GTA) included in the trip flare's shipping box that gives detailed instructions for installing the flare.

4-42. While trip flares are primarily mounted, they can also be hand-thrown.

MOUNTED

4-43. Soldiers should mount the trip flare using the following procedures:

(1) Choose a location.

NOTE: The location chosen for the flare should be to the right (looking toward the enemy) of the field to be illuminated, so the trip wire, when attached, runs to the right of the flare (when facing the trigger).

WARNING

Surface trip flares can cause fires when mounted on dry tender.

---

(2) Using two of the nails supplied, nail the mounting bracket (with ends of the two tabs upward) to a stake, post, or suitable support at the height desired for the trip wire (usually 15 to 18 inches above the ground).

```
WARNING

Never mount a surface trip flare above knee level.
```

(3) Mount the flare by sliding the two square holes of the anchor clip over the mating tabs on the holder, and press the flare down until it is locked in position.

NOTE: If desired, a third nail may be driven through the hole in the lower end of the anchor clip.

(4) Fasten one end of the trip wire to the post, stake, or other rigid object at the desired distance from the flare (usually about 40 feet) and at the right of the flare (when facing the flare trigger).

(5) Press the fuze safety lever down with one hand, and rotate the trigger one-quarter turn counterclockwise against the spring pressure with the other hand to the vertical position, so the lower end of the safety lever is behind the upper end of the trigger.

(6) Pull the loose end of the trip wire taut, and fasten it to the hole in the lower end of the trigger.

(7) Check to see that the trip wire is taut and fastened at both ends, and that the trigger is vertical with the fuze safety lever (behind the upper end of the trigger) so when the pull ring and safety pin are withdrawn, the safety lever is still held by the trigger.

```
DANGER

ENSURE THAT THE TRIGGER IS VERTICAL WITH THE FUZE SAFETY
LEVER (BEHIND THE UPPER END OF THE TRIGGER) SO WHEN THE
PULL RING AND SAFETY PIN ARE WITHDRAWN, THE SAFETY LEVER
IS STILL HELD BY THE TRIGGER. FAILURE TO DO SO CAN RESULT
IN ACCIDENTAL IGNITION OF THE TRIP FLARE.
```

(8) Hold the lever with one hand, while carefully withdrawing the pull ring and safety pin from the flare.

(9) Carefully release the hold on the safety lever, while making sure the lever is held in place by the upper end of the trigger.

(10) Move to a safe location.

```
WARNINGS

The minimum safe distance from an ignited surface trip flare is 2
meters because of sparks and the popping of burning magnesium.

Never look directly at a burning surface trip flare. The intense
flame can injure your eyes.

At close ranges, surface trip flares may damage night vision
devices and sights.
```

HAND-THROWN

4-44. To throw a trip flare by hand—

(1) Remove the flare from the mounting bracket.

(2) Follow the same preparation, gripping, and throwing procedures as those for a grenade.

NOTE: See Chapter 3 for more information about grenade preparation, gripping, and throwing procedures.

WARNINGS

Surface trip flares can cause fires when thrown on dry tender.

The minimum safe distance from an ignited surface trip flare is 2 meters because of sparks and the popping of burning magnesium.

Never look directly at a burning surface trip flare. The intense flame can injure your eyes.

At close ranges, surface trip flares may damage night vision devices and sights.

DO NOT attempt to cook off a trip flare. The fuze has a .0-second time delay.

MAINTENANCE

4-45. When exposed to the environment, trip flares require PMCS. If not removed, dirt or sand can cause the trip flare to malfunction.

CLEANING

4-46. To clean the trip flare—
- Wipe the dirt off the flare using a clean, dry, lint-free cloth.
- Use a fine-bristled camel hair brush to remove any foreign matter or debris.

REMOVAL

4-47. To remove a trip flare—
(1) Carefully depress the safety lever to align the holes in the lever and the fuze.
(2) Insert the safety pin.
(3) Detach the trip wire from the trigger, while holding the safety lever against the flare.
(4) Rotate the trigger to its original position.
(5) Remove the nails from the mounting bracket and the anchor clip.
(6) Return the flare to its original position and packing.

SECTION III. SIMULATED SIGNALS

Some pyrotechnic simulators can be used to provide early warning signals and to illuminate the immediate area; however, they are primarily designed to imitate the sounds and effects of combat detonations during field training exercises.

EARLY WARNING SIMULATORS

4-48. Early warning simulators generate different effects upon initiation, but are used the same way; they are activated by triggering trip wires attached to the igniter cords.

INSPECTION

WARNING

Handheld communication signals must be inspected before use and properly secured to avoid serious injury.

Initial Inspection

CAUTION

During storage, keep boxes sealed. Duds or improper functioning could occur if exposed to moisture for long periods of time. Open just before use.

4-49. When in bulk, communication signals are secured in shipping containers (Figure 4-11). Personnel should inspect the shipping container upon receipt. Shipping containers that are damaged should not be opened; they should be returned to the ASP or disposed of using the methods outlined in the unit SOP.

4-50. Within the shipping container are barrier bags. Upon removing the sealed barrier bags from the shipping container (Figure 4-12), personnel should inspect each barrier bag and identify any of the following discrepancies:

- The barrier bag has been damaged.
- The seal on the barrier bag shows signs of tampering.

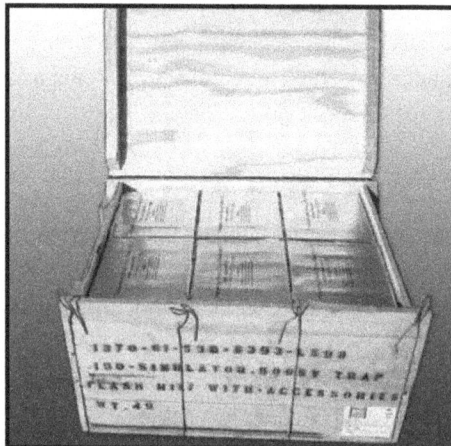

Figure 4-12. Early warning simulator shipping container and barrier bags.

4-51. Inside the barrier bags are cardboard shipping boxes. Upon removing the cardboard shipping boxes from the sealed barrier bags (Figure 4-13), personnel should inspect each shipping box and identify any of the following discrepancies:

- The shipping box has been damaged.
- The shipping box shows signs of tampering.

Figure 4-13. Early warning simulator cardboard shipping box.

4-52. Then, personnel should remove the early warning simulator from the container. Once the signal has been removed from the container (Figure 4-14), personnel should inspect the signal and identify any of the following discrepancies:

- Holes are visible in the body.
- The seal securing the cap is broken.
- The mounting bracket is damaged.
- The safety clip is damaged or not present.

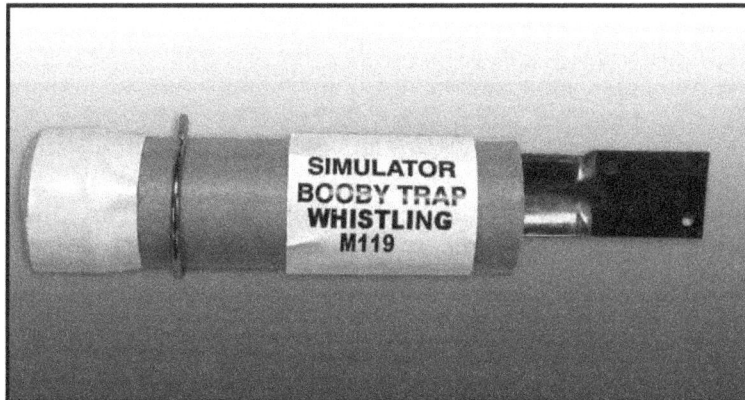

Figure 4-14. Early warning simulator removed from the cardboard shipping box.

NOTE: If any of the discrepancies are found upon receipt of newly issued early warning simulators, personnel should return the simulator and shipping containers to the issuing person or dispose of it in accordance with the unit SOP.

Before Storing

4-53. Before securing early warning simulators, personnel should take the following safety precautions:

- Ensure the body is not bent or punctured.
- The seal securing the cap is not broken.
- The safety clip is present and not damaged.
- All other kit components (Figure 4-15) are present.

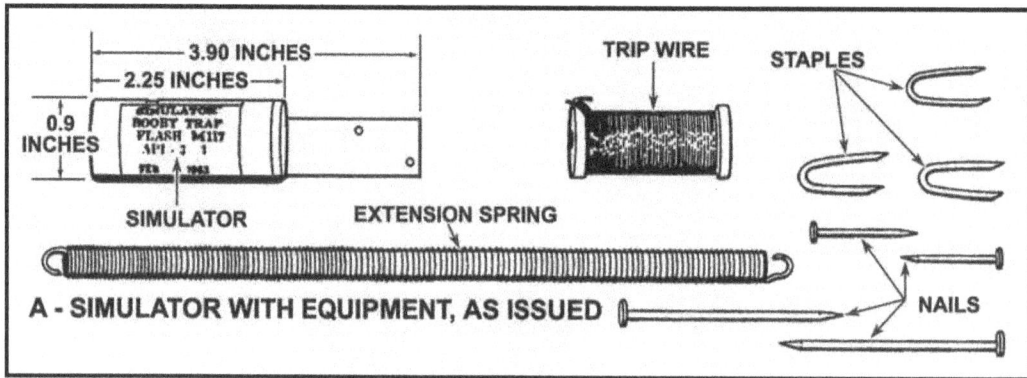

Figure 4-15. Kit components.

STORAGE

4-54. Personnel should carry early warning simulators in accordance with the unit SOP. When carrying early warning simulators, personnel should adhere to the following guidelines:

- Ensure that the early warning simulator is placed in a secure, dry area.

> **CAUTION**
>
> Duds or improper functioning could occur if exposed to moisture for long periods of time. Open just before use.

- Do not put adhesive tape around any portion of the early warning simulator during storage.

> **WARNING**
>
> **Do not bend, tamper, modify or otherwise alter a trip flare. Do not tape any portion of the trip flare during storage.**

USE

> **DANGER**
>
> EARLY WARNING SIMULATORS MUST BE MOUNTED; DO NOT ACTIVATE THEM BY HAND. EARLY WARNING SIMULATORS WILL IMMEDIATELY ACTIVATE.
>
> EARLY WARNING SIMULATORS WILL SERIOUSLY INJURE PERSONNEL WITHIN 2 FEET.
>
> NEVER OPEN A SIMULATOR; THE PHOTOFLASH POWDER IS EXTREMELY SUSCEPTIBLE TO FLASH IGNITION BY EVEN A SLIGHT AMOUNT OF FRICTION.

```
┌─────────────────────────────────────────────────────────────────┐
│                          WARNINGS                                 │
│                                                                   │
│   Early warning simulators must not be activated in loose gravel, │
│   sticks, or other materials that could become projectiles, nor   │
│   should they be thrown into dry leaves, grass, or other flammable│
│   materials. Dry grass or leaves within 3 feet may become ignited.│
│                                                                   │
│   DO NOT tape or wire early warning simulators to any surface. Use│
│   nails.                                                          │
│                                                                   │
│   DO NOT remove the simulator cap before use.                     │
└─────────────────────────────────────────────────────────────────┘
```

NOTE: There is a GTA included in the early warning simulator's cardboard shipping box that gives detailed instructions for installing the simulator. The same illustrative instructions may also be found in TM 9-1370-207-10.

4-55. The instructions included with the early warning simulator show how to install the simulator on a tree. This is just one technique of installing the early warning simulator.

4-56. To install an early warning simulator—

(1) Select two objects, such as trees or stakes, not more than 20 feet apart.
(2) About 6 inches above the ground, drive a large nail into one object.
(3) Drive a staple about 2 inches above and to the right of the nail.
(4) Drive a second staple into the object about 20 inches above the first staple.
(5) Drive a large nail 1 inch below the top staple.

NOTE: This will be used to temporarily hold the spring on the lower staple.

(6) Extend the spring to the nail driven in during Step 5.
(7) Make a 6-inch loop in one end of the tripwire, and tie it with a double-knot. Thread the loop down through the top staple, and attach to the upper end of the spring only.
(8) Maintain tension on the tripwire, and run it down the object, under the bottom nail, and toward the second object.
(9) Drive a staple into the second object at the lowest point that will allow free travel of the tripwire.
(10) Maintain tension on the tripwire, and tie the wire to a large nail, just below its head. Wedge the nail between the staple and the object to ensure a taut and secure tripwire.
(11) Carefully unhook the extended spring from the temporary nail. The spring should keep the wire taut.
(12) Remove the tape securing the cap. Remove the cap from the simulator, and allow the pull cord to hand freely.
(13) Nail the simulator about 4 inches above the top of the spring.
(14) Leave 1 or 2 inches of slack in dangling cord, and tie the cord to the end of the spring that is fastened to the tripwire.

Maintenance

4-57. There is no unit level maintenance for the early warning simulators. Turn in unused items to ammunition support area as soon as possible, or dispose of in accordance with unit SOP. Provide as much protection for these items by repacking in original containers, if available, or equivalent improvised packaging.

- All repacking should be tightly wrapped, clearly marked and waterproof.
- Avoid exposure to moisture and rough physical contact.

DESTRUCTION PROCEDURES

4-58. Selection of the method of destruction requires imagination and resourcefulness in the utilization of the facilities at hand under the existing conditions. In general, destruction of handheld signals can be accomplished most effectively by burning or firing, or a combination of these methods.

GROUND-BURST SIMULATOR

4-59. The M115A2, projectile ground-burst simulator replicates the detonation of artillery and mortar projectiles or artillery-type rockets (Figure 4-16). It is activated by pulling its M3A1 friction delay igniter cord and immediately thrown into a cleared area. After a 6- to 10-second delay, it produces a high-pitched whistle that lasts 2 to 4 seconds and then detonates with a loud report and brilliant flash.

NOTE: Instructions for the ground-burst simulator are printed directly on the simulators.

WARNING

The M115A2 projectile ground-burst simulator must not be used near personnel due to potential hazard from fragmentation. Ensure the simulator is not thrown to any point within 35 meters of unprotected personnel. When using the M115A2 ground-burst simulator, the thrower should turn away from the simulator after throwing.

WARNING

For protection, personnel throwing the M115A2 simulator must wear the following items: ear protection, safety eyewear, and a protective helmet and vest. The user must wear a standard-issue leather glove on the throwing hand.

WARNING

The M115A2 must not be activated in loose gravel, sticks, or other materials that could become projectiles, nor should they be thrown into dry leaves, grass, or other flammable materials.

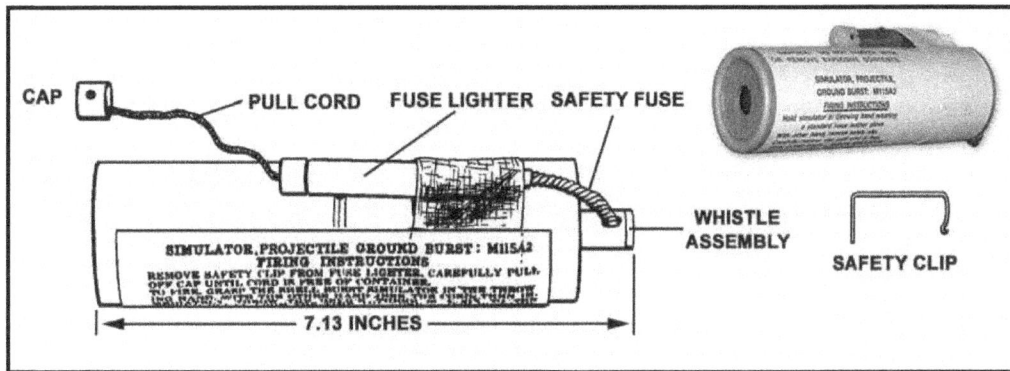

Figure 4-16. M115A2 ground-burst simulator.

INSPECTION

4-60. Inspection at unit level consists of a visual check of packaging materials. Do not open any moisture-proof container or barrier bag because the item must be protected from moisture just prior to use.

4-61. The most commonly encountered packaging defects are listed below:
- Outer containers (boxes) damaged, weathered, or rotted to the extent that contents are not protected.
- Inner container damaged to the extent that contents are not protected or cannot be readily removed.
- Container cap or closure not secured to the extent that contents are not protected.
- Inner containers wet (except metal), rusted, moldy, or mildewed.
- Hardware or banding loose, missing, broken, or ineffective.
- Handle or cleat missing or broken.
- Contents loose to the extent that item may be damaged in handling.

STORAGE

4-62. Take the following precautions when storing M115A2 ground-burst simulators—
- Select level, well-drained sites free from readily ignitable and flammable materials.
- Provide nonflammable or fire-resistant overhead covers (e.g., tarpaulin) for all items. Maintain overhead space of approximately 18 inches (46 centimeters) between cover and items. Keep cover at least 6 inches (15.5 centimeters) from pile on the ends and at sides to permit circulation of air.
- Temporarily store unserviceable items in a segregated area.

USE

4-63. To prepare and throw a ground-burst simulator by hand—
- Remove safety clip from fuze lighter.
- Grasp simulator in throwing hand. Carefully remove cap until free end of igniter and cord is partially extended.
- Assume throwing position (standing, kneeling, or prone). Jerk the pull-cord once, and throw immediately.

NOTE: See Chapter 3 for throwing positions.

MAINTENANCE

4-64. There is no unit level maintenance for the ground-burst simulators. Turn in unused items to ammunition support area as soon as possible. Provide as much protection for these items by repacking in original containers, if available, or equivalent improvised packaging.

- All repacking should be tightly wrapped, clearly marked and waterproof.
- Avoid exposure to moisture and rough physical contact.

DESTRUCTION PROCEDURES

4-65. Selection of the method of destruction requires imagination and resourcefulness in the utilization of the facilities at hand under the existing conditions. In general, destruction of handheld signals can be accomplished most effectively by burning or firing, or a combination of these methods.

HAND GRENADE SIMULATOR

4-66. The M116A1 hand grenade simulator (Figure 4-17) is used to create battle noises and flashes during training. It differs from the ground-burst simulator in that it is shorter and does not emit a high-pitched whistle before detonation. The hand grenade simulator is thrown in the same manner as a live grenade. It creates a flash and loud report 5 to 10 seconds after ignition.

WARNINGS

The safety radius for the M116A1 simulator is 15 meters. If this distance is not observed, hearing damage and possible fragmentation injury could result. The M116A1 projectile ground-burst simulator must not be used near personnel due to potential hazard from fragmentation.

The M116A1 must not be activated in loose gravel, sticks, or other materials that could become projectiles, nor should they be thrown into dry leaves, grass, or other flammable materials.

For protection, personnel throwing the M116A1 simulator must wear the following items: ear protection, safety eyewear, and a protective helmet and vest. The user must wear a standard-issue leather glove on the throwing hand.

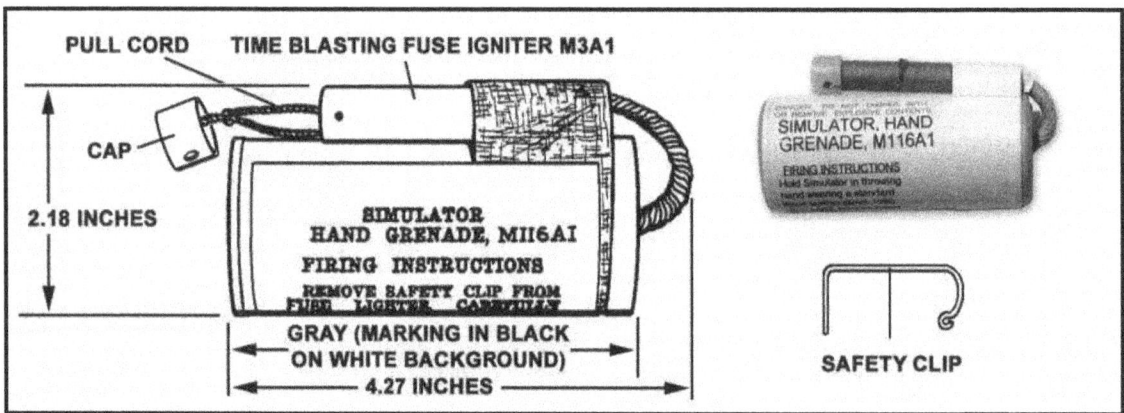

Figure 4-17. M116A1 hand grenade simulator.

NOTE: Instructions for the hand grenade simulator are printed directly on the simulators.

SECTION IV. ILLUMINATION GROUND SIGNAL KITS

The pen gun flare supports the small-unit leader in fire control, maneuver, and initiating operations such as ambushes. These signals are also a component of air crewmen's survival vest and are used for distress signaling or to identify ground locations for aircraft (Figure 4-18).

WARNINGS

At close-range, these signals can injure or kill if they strike a person.

When signaling an aircraft, do not aim directly at the aircraft; the signals, regardless of color, may appear to be small arms fire.

While the flares are small and usually burn out before reaching the ground, they can ignite fires.

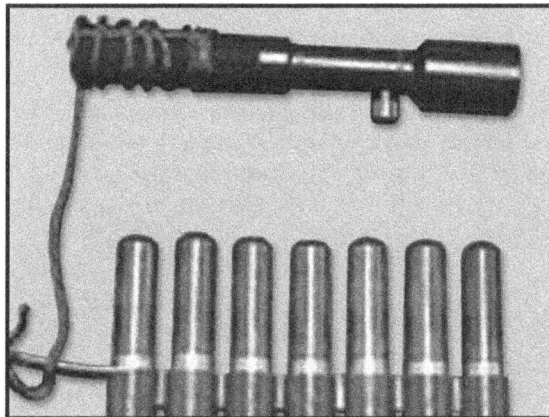

Figure 4-18. M260 illumination ground signal kit.

INSPECTION

4-67. Inspection at unit level consists of a visual check of packaging materials. Do not open any moisture-proof container or barrier bag because the item must be protected from moisture just prior to use.

4-68. The most commonly encountered packaging defects are listed below:
- Outer containers (boxes) damaged, weathered, or rotted to the extent that contents are not protected.
- Inner container damaged to the extent that contents are not protected or cannot be readily removed.
- Container cap or closure not secured to the extent that contents are not protected.
- Inner containers wet (except metal), rusted, moldy, or mildewed.
- Hardware or banding loose, missing, broken, or ineffective.
- Handle or cleat missing or broken.
- Contents loose to the extent that item may be damaged in handling.

STORAGE

Take the following precautions when storing the illumination ground signal kit—

- Select level, well-drained sites free from readily ignitable and flammable materials.
- Provide nonflammable or fire-resistant overhead covers (e.g., tarpaulin) for all items. Keep cover at least 6 inches (15.5 centimeters) from pile on the ends and at sides to permit circulation of air.
- Temporarily store unserviceable items in a segregated area.

USE

4-69. To operate an illumination ground signal kit—

(1) Select the signal to be fired by color (if using the M186 pen flare kit). If the bandoleer contains more than one signal of the chosen color, use the one farthest from the lanyard.

(2) Remove and discard the plastic cap (M185 and M186 only).

(3) Cock the projector by moving the trigger to the safety slot (M185 and M186 only).

(4) Carefully thread the projector onto the signal. Take care not to dislodge the trigger from the safety slot (M185 and M186 only).

(5) Aim in the chosen direction.

(6) Fire by moving the trigger to the bottom of the slot and releasing it with a snap.

(7) If the expended signal is on the end of the bandoleer or if the signals between the expended signal and the end have been used, cut the bandoleer and discard the waste.

(8) Return the partly used kit to the carrier bag, and seal with tape.

MAINTENANCE

4-70. There is no unit level maintenance for the illumination ground signal kit. Turn in unused items to ammunition support area as soon as possible. Provide as much protection for these items by repacking in original containers, if available, or equivalent improvised packaging.

- All repacking should be tightly wrapped, clearly marked and waterproof.
- Avoid exposure to moisture and rough physical contact.

DESTRUCTION PROCEDURES

4-71. Selection of the method of destruction requires imagination and resourcefulness in the utilization of the facilities at hand under the existing conditions. In general, destruction of handheld signals can be accomplished most effectively by burning or firing, or a combination of these methods.

Chapter 5

Employment Considerations

Soldiers employ hand grenades and pyrotechnics throughout the spectrum of warfare to conceal positions and to inflict greater casualties.

APPLICATION

5-1. Hand grenades and pyrotechnics provide the individual Soldier with a number of highly versatile and effective weapon and signaling systems. Each system is designed for a specific application (Tables 5-1 and 5-2).

Table 5-1. Types of hand grenades and their applications.

TYPE	APPLICATION
Fragmentation hand grenades	Supplement small arms fire against enemies in close combat
	Destroy or disable equipment
Offensive hand grenades (concussion grenades)	Destroy above and below ground manmade structures
	Kill or stun enemy divers during waterborne operations
Stun hand grenades	Confuse, disorient, or momentarily distract a potential threat in forced entry scenarios, selected urban operations, or crowd control operations
Incendiary hand grenades	Immobilize vehicles
	Destroy equipment and weapon systems
	Destroy munitions and start fires

Table 5-2. Types of pyrotechnic signals and their applications.

TYPE	APPLICATION
Colored smoke grenades	Identify places or objects
	Mark positions
White smoke grenades	Conceal
	Conceal or create a smoke screen
Riot-control hand grenades	Control crowds or riots
Star clusters, star parachutes, and smoke parachutes	Signal
	Illuminate
Surface trip flares	Destroy equipment
	Provide early warning
	Illuminate an immediate area
Booby trap simulators (flash, illuminating, and whistling)	Early warning signals
Ground-burst simulators and grenade simulators	Simulate battle noises and battlefield effects (i.e., shells in flight and ground-burst explosions)

RULES OF ENGAGEMENT

5-2.   To properly employ any type of hand grenade or pyrotechnic signal, Soldiers must know the—
*   Characteristics and capabilities of the chosen grenade or pyrotechnic.
*   Location of all friendly forces.
*   Sector of fire.
*   Projected arc or path of the grenade or pyrotechnic, ensuring that it is free of obstacles.
*   Direction of the wind (when employing handheld and ground smoke signals).

5-3.   Further, Soldiers should use the buddy or team system when employing grenades and pyrotechnics. For example, one Soldier can provide covering fire, or as a combined team, Soldiers can employ grenades on the target and ground smoke within the target area.

CONSIDERATIONS

5-4.   When employing any type of hand grenade or pyrotechnic signal, Soldiers must ask themselves the following questions:
*   What types of grenades do the rules of engagement (ROE) permit and restrict?
*   What effect is desired (e.g., kill, stun, obscure, destroy equipment, mark a location, etc.)?
*   Does the structural integrity of the room and building permit the types of grenades selected for use?

WARNING

DO NOT use fragmentation or concussion grenades in buildings that have walls of thin veneer material. Fragmentation grenade particles can penetrate partitioned walls, and concussion grenades can weaken the structure of the building or cause portions of the building to collapse inward.

---

*   Will the scheme of maneuver permit the use of fragmentation grenades and not cause fratricide?
*   Will the type of grenade used cause a fire in an undesired location?

OFFENSIVE OPERATIONS

5-5.   The fragmentation hand grenade is the primary type of grenade used during offensive operations; however, offensive operations can also involve the use of offensive and stun hand grenades. These operations include clearing—
*   Confined spaces.
*   Trenches.
*   Bunkers.
*   Rooms.
*   Entrenched positions.

CONFINED SPACES

5-6.   If the enemy is located in a confined space, such as a bunker, building, or fortified area, the offensive grenade or stun grenade may be more appropriate than the fragmentation hand grenade. These types of grenades are much less lethal than fragmentation grenades on an enemy in the open, but they are very effective against an enemy within a confined space and are safer to employ in the confines of a smaller space. Selection of a grenade, however, depends upon availability and mission analysis.

WARNING

The concussion produced by offensive grenades in enclosed areas is greater than those produced by the fragmentation grenade. This may weaken or cause collapse of a structured foundation.

TRENCHES

5-7.   A mix of fragmentation grenades and offensive grenades should be used to clear enemy fortified trenches: fragmentation grenades to gain and clear the enemy's trench lines, and offensive grenades to clear and destroy any fortified positions.

NOTE: See FM 3-21.8 and Battle Drill 07-8-1 for detailed instruction on how a squad enters and clears a trench with the use of grenades.

WARNING

To ensure the safety of squad members during tactical operations, Soldiers employing fragmentation grenades must shout "FRAG OUT" before throwing the grenade.

CLEARING BUNKERS

5-8.   A mix of fragmentation and offensive grenades can be used to clear bunkers: fragmentation grenades to suppress enemy fires during movement, and a mix of fragmentation and offensive grenades to destroy fortified positions.

NOTE: See FM 3-21.8 and Battle Drill 07-8-1 for detailed instruction on how a squad clears a bunker with the use of grenades.

WARNING

To ensure the safety of squad members during tactical operations, Soldiers employing fragmentation grenades must shout "FRAG OUT" before throwing the grenade.

CLEARING A ROOM

5-9.   The following is an example of how a squad clears a room using hand grenades:

WARNING

Conduct a structural analysis before using fragmentation or concussion grenades to clear rooms of a building. Grenade fragments can penetrate walls and cause injury to the clearing team. Concussion grenades may weaken the structure, causing part or all of the building to collapse on the clearing team.

---

(1) The squad leader and assaulting fire team approach the room and position themselves at either side of the entrance.

(2) A Soldier of the assaulting fire team cooks off a fragmentation or concussion grenade (2 seconds maximum), shouts "FRAG OUT" or "CONCUSSION OUT" to alert friendly personnel, and then throws the grenade into the room.

CAUTION

If a stun grenade is chosen for room clearing, do not cook off the grenade. The stun grenade has a short time fuze; the thrower will not have adequate time to dispose of the grenade before it explodes. This can cause serious injury to the thrower and friendly personnel nearby.

NOTE: If stealth is a factor, the thrower alerts the team/squad of the type of grenade by holding it up so that each member can see the grenade and take appropriate safeguard actions.

DANGER

SOLDIERS MUST BE CONSCIOUS OF FEATURES WITHIN THE ROOM TO BE CLEARED. GRENADES TEND TO ROLL DOWN STAIRS. THIS CAN NULLIFY THE DESIRED EFFECT(S) OR CAUSE FRIENDLY CASUALTIES.

---

(3) After the grenade explodes, the lead man in the clearing team enters the room, eliminates any immediate threat, and moves to his point of domination.

(4) The remainder of the clearing team enters the room and moves to their points of domination, eliminating any threat.

(5) Once cleared, the team marks the room in accordance with the unit SOP.

NOTE: See FM 3-21.8 and Battle Drill 07-8-1 for detailed instruction on how a squad clears a room with the use of grenades.

ENTRENCHED POSITIONS

5-10. Against Soldiers in open trenches or fighting positions, Soldiers should throw a fragmentation grenade or an offensive concussion grenade so that it bursts over the target. If the targets are on sloping ground, Soldiers should use above-ground detonation to prevent the grenade from rolling away from the target before detonating. Above-ground detonation also prevents the enemy from securing the grenade and throwing it back within the 4- to 5-second fuze delay.

NOTE: See Chapter 3 for more information about above-ground detonation (cooking off).

DEFENSIVE OPERATIONS

5-11. The fragmentation grenade is the primary hand grenade used in defensive operations. It can be used in conjunction with other weapons to destroy remnants of an attacking enemy force that may succeed in penetrating the more distant barriers and final protective fires. The fragmentation hand grenade further disrupts the continuity of the enemy attack, demoralizes the enemy soldier, and forces the enemy into areas covered by direct-fire weapons such as rifle and machine gun fire and Claymore mines. Using fragmentation hand grenades on dismounted enemy forces at a critical moment in the assault can be the final blow in taking the initiative away from the enemy.

NOTE: See FM 3-21.8 for more information regarding the use of grenades in defensive operations.

DEFENSE FROM INDIVIDUAL FIGHTING POSITIONS

5-12. From individual fighting positions, fragmentation hand grenades can be used to cover close-in dead space. Soldiers should use these grenades in conjunction with ground flares along the front of their defensive positions. Potential avenues of approach through the unit's perimeter should be marked with a reference to identify them as primary hand grenade targets.

5-13. The following rules apply when employing fragmentation hand grenades from fighting positions:

- Clear overhead obstructions that may interfere with the path of the thrown grenade. Do this at the same time direct-fire fields of fire are cleared.
- Rehearse grenade employment; know where the primary target is located.
- Keep 50 percent of the fragmentation grenades at the ready in the fighting position, leaving the remaining fragmentation grenades on the LCE/ETLBV.
- Rehearse actions needed if an enemy grenade lands in the fighting position.
- Employ fragmentation hand grenades against enemy soldiers located in defilade positions as first priority. This lessens the danger to friendly Soldiers and helps cover terrain not covered by direct-fire weapons.
- Reconnoiter the alternate and supplementary positions, and determine the priority for the fragmentation hand grenade target.

WARNING

Grenade fuze time delays can vary. If an enemy grenade lands in the position, immediately evacuate the fighting position, or if nearby, quickly lay flat on the ground.

RETROGRADE OPERATIONS

5-14. Most of the employment considerations applicable to the use of hand grenades and pyrotechnics in the defense are equally applicable to retrograde operations. Considerations unique to retrograde operations relate to creating obstacles, marking friendly force locations, breaking contact, and communicating.

CREATE OBSTACLES

5-15. When terrain conditions permit, Soldiers can use incendiary grenades to impede and disrupt enemy movement by initiating fires in specific areas.

MARK LOCATIONS

5-16. Soldiers can use prearranged smoke colors to mark friendly force positions and identify friendly forces.

BREAK CONTACT

5-17. During retrograde operations, some elements of the friendly force may become decisively engaged. Soldiers can use fragmentation, white smoke, and CS grenades to break contact and regain flexibility of maneuver. Use of hand grenades in volley fire following the employment of white smoke is especially effective; the smoke obscures enemy observation of friendly force movement from covered positions, and the fragmentation grenades force the enemy to cover.

COMMUNICATIONS

5-18. Soldiers can use prearranged handheld pyrotechnic signals to communicate friendly force movements.

URBAN OPERATIONS

5-19. By definition, urban areas house large quantities of people and contain large numbers of buildings. The enemy may be intermingled with noncombatants, and collateral damage must be limited. Because of these factors, the ROE may be more restrictive than under other combat conditions. Table 5-3 outlines the guidelines for hand grenade employment.

NOTE: See FM 3-06.11 for use of grenades and pyrotechnics in urban areas.

Table 5-3. Hand grenade employment during urban operations.

TYPE OF HAND GRENADE	EMPLOYMENT
Stun-type grenades	Use when noncombatants and friendly forces may be intermingled with threat forces. Throw into rooms prior to entering to cause confusion and hesitation (especially useful if the structural integrity of the building does not permit the use of fragmentation or concussion grenades).
Riot-control grenades	Use during urban operations to maintain control. Should be employed only when command-directed.
Fragmentation grenades	Throw at assaulting enemy troops between buildings or on streets from windows, doors, or manmade apertures (mouseholes). Should be employed only when command-directed.
Smoke grenades	The M106 provides a near instantaneous screen of dense smoke and is safe to use inside of urban structures, subterranean locations, and caves. The M106 should be used in lieu of the AN-M8 HC and the M83 TA white smoke hand grenade when inside of confined spaces, and when encountering enemy in close quarters.

WARNING

Use of the AN-M8 HC and the M83 TA white smoke hand grenade are harmful to personnel and may cause fires inside of confined spaces. These smoke grenades inside buildings may displace oxygen in poorly ventilated rooms and make breathing difficult while also rendering protective masks ineffective.

AIR OPERATIONS

WARNING

Do not throw fragmentation grenades from low-flying or hovering helicopters. The fragments present a hazard to the aircraft and its passengers.

5-20. Generally, throwing hand grenades from medium or high-flying helicopters is limited to mission-critical situations.

USE UNDER ADVERSE CONDITIONS

5-21. While hand grenade procedures do not change when employed under adverse conditions, special precautions must be taken.

MOPP4

5-22. Wearing gloves, especially gloves used during MOPP, inhibits the thrower's feel and could decrease his throwing ability and range. The thrower should execute arming and throwing procedures carefully and concentrate on using the proper grip. Observing each arming action (removal of safety clip and safety pin) is also recommended in MOPP.

LIMITED VISIBILITY

5-23. Depth perception is generally impaired under limited visibility conditions. Throwers must have clear fields of fire with no overhead obstructions.

This page is intentionally left blank.

Appendix A

Characteristics of Grenades

There are five types of hand grenades: fragmentation, chemical, offensive, nonlethal, and practice. This appendix outlines the characteristics of these hand grenades.

NOTE: A confidence clip (Figure A-1) has been introduced to hand grenades equipped with a safety pin and pull ring.

Figure A-1. Confidence clip.

FRAGMENTATION HAND GRENADES

A-1.   Fragmentation hand grenades produce high-velocity projection of fragments.

M67 FRAGMENTATION HAND GRENADE

A-2.   The M67 (Figure A-2) is the most commonly used fragmentation hand grenade. Table A-1 outlines the grenade's components and characteristics in accordance with TM 9-1330-200-12.

Figure A-2. M67 fragmentation hand grenade.

Table A-1. Components and characteristics of M67 fragmentation grenade.

COMPONENTS AND CHARACTERISTICS	DETAILS
Body	Steel sphere with a scored steel spring for fragmentation
Filler	6.5 ounces of Composition B
Fuze	M213
Safety Features	Safety clip Safety pin and pull ring with confidence clip Safety lever
Fuze Delay	4 to 5.5 seconds
Total Weight	14 ounces
Throwing Distance of Average Soldier	35 meters
Effective Casualty-Producing Radius	15 meters
Killing Radius	5 meters
Colors and Markings	Olive drab body with yellow markings

WARNING

Although the killing radius of the M67 grenade is 5 meters and the casualty-producing radius is 15 meters, fragments can disperse as far as 230 meters.

PRACTICE HAND GRENADES

A-3.   Practice hand grenades simulate the effects of other hand grenades so that training can be conducted with a reduced chance of injury to personnel or damage to property.

M69 PRACTICE HAND GRENADE

A-4.   The M69 practice hand grenade is used for all individual and collective training tasks. The M69 practice hand grenade (Figure A-3) provides realistic training and familiarizes the Soldier with the functioning and characteristics of the M67 fragmentation hand grenade. Table A-2 outlines the grenade's components and characteristics.

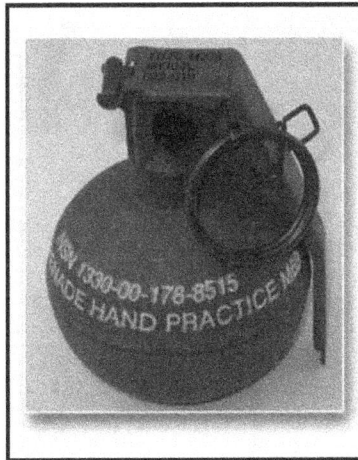

Figure A-3. M69 practice hand grenade.

Table A-2. Components and characteristics of M69 practice grenade.

COMPONENTS AND CHARACTERISTICS	DETAILS
Body	Hollow steel sphere
Filler	None
Fuze	M228, which is inserted into the grenade body
Safety Features	Safety clip Safety pin and pull ring with confidence clip Safety lever
Total Weight	14 ounces
Throwing Distance of Average Soldier	40 meters
Fuze Delay	4 to 5.5 seconds
Effects	Small puff of white smoke and a loud popping noise
Colors and Markings	Light blue with white markings; the safety lever of the fuze is light blue with black markings and a brown tip

---

WARNING

Fuze fragments may exit the hole in the base of the grenade body and cause injuries.

NOTE: The grenade body can be used repeatedly by replacing the fuze assembly.

M228 Detonating Fuze

A-5. During practice events and for qualification, each Soldier is required to throw several M69 practice hand grenades armed with the M228 detonating fuze (Figure A-4). Although it takes only about a minute or less to install or replace a used fuze, a company-size element will use several hundred; preparing practice grenades for all participants is not feasible. Soldiers should be given instruction on installing and removing a fired M228 fuze.

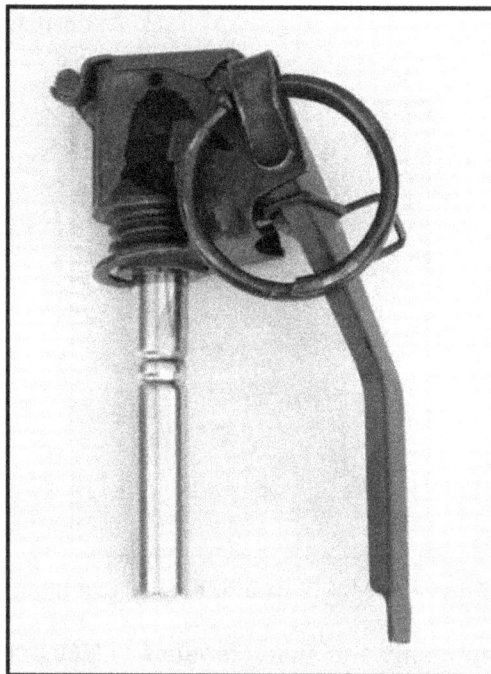

Figure A-4. M228 detonating fuze.

Install

NOTE: The M228 fuze for the M69 practice grenade will come with the confidence clip attached. However, the confidence clip can be issued separately and then installed on the M228 fuze before connecting the fuze to the M69 practice grenade.

A-6. To install the M228 fuze with confidence clip in the M69 grenade—

> **CAUTION**
>
> Do not hold or touch the fuze igniter. The igniter, made of light aluminum, has an explosive charge that could be damaged if twisted, and could cause injury to the hand.

(1) Hold the head of the M228 fuze with the fingers of the throwing hand and the confidence clip in the opposite hand (Figure A-5).

Figure A-5. Gripping the M228 detonating fuze.

(2) Place the circular opening of the confidence clip over the threaded end of the M228 fuze with the fingers of the opposite hand, ensuring the clip end is up and on the same side as the safety pin and pull ring (Figure A-6).

Figure A-6. Placing the confidence clip over the threaded end of the M228 fuze.

(3) Grip the M69 body with the non-throwing hand, taking care to keep any part of the hand away from the firing port (non-threaded end) of the grenade (Figure A-7). While holding the body of the M69, turn the grenade so that the threaded end is facing inward and the firing port of the grenade is facing away from your body.

(4) Insert the M228 fuze into the threaded end of the M69, and turn the fuze or grenade body clockwise until the fuze and confidence clip is secure.

WARNING

Keep all portions of the hand away from the M69 practice grenade firing port (non-threaded end) when installing the M228 detonating fuze and when throwing the grenade. The fuze explosive charge can cause injury to the hand and fingers when it exits through the firing port.

CAUTION

The M228 practice fuze should be only finger-tight. Do not over-tighten the fuze. This could damage threads in the M69 body and the threaded end of the fuze.

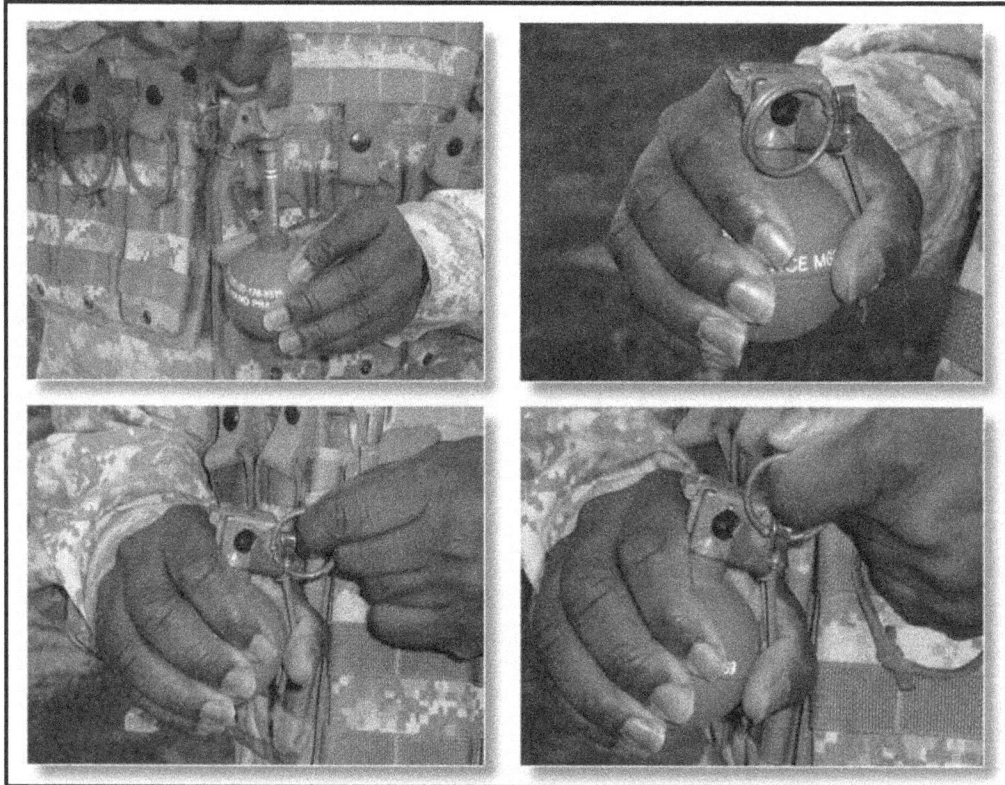

Figure A-7. Insert the M228 fuze into the M69 practice hand grenade body and secure pull ring to the confidence clip (right hand).

(5) With the index finger inserted in the pull ring, rotate the pull ring around until the single loop portion is within the confidence clip area. Align the pull ring under the confidence clip (Figure A-8).

(6) Pull up on the pull ring until it snaps under the confidence clip (Figure A-8).

WARNING

When attaching a confidence clip to a M228 detonating fuze pull ring, do not remove or dislodge the safety pin. Ensure the safety pin is spread open in a duckbill shape to prevent the safety pin from falling out. If the safety pin comes out, do not attempt to put the safety pin back in. Immediately alert any personnel in the area, throw the fuze in a safe area, and take the same actions as you would for a live hand grenade.

NOTE: If Soldier is a left-hand thrower, he will insert the fuze in the same manner as described above. See Figure A-7 for clarification.

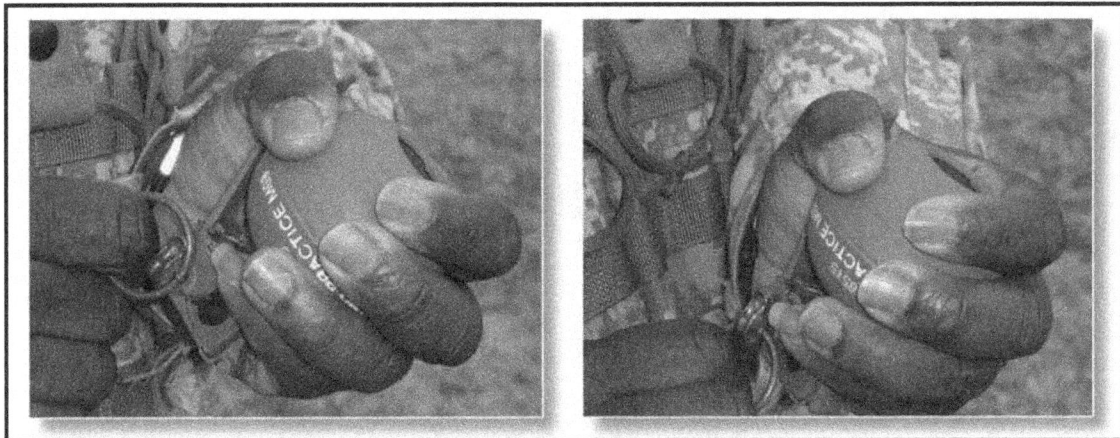

Figure A-8. Securing the pull ring to the confidence clip (left hand).

Remove

A-7. To remove a fired M228 detonating fuze—

WARNING

Keep all portions of the hand away from the M69 practice grenade firing port (non-threaded end) when removing a fired M228 detonating fuze. Residue from an exploded fuze can cause lacerations to the hand and fingers.

---

(1) Grip the M69 body with the non-throwing hand, taking care to keep any part of the hand away from the firing port (non-threaded end) of the grenade. While holding the M69 grenade body, turn the grenade so that the head of the fuze is facing inward and the firing port of the grenade is facing away from your body.

(2) Grip the head of the M228 fuze with the fingers of the throwing hand.

WARNING

Keep all portions of the hand and fingers away from the lower end of the M228 detonating fuze after removing a fired fuze. Jagged edges of the exploded fuze end can cause lacerations to the hand and fingers.

---

(3) Turn the M228 fuze or grenade body counterclockwise until the fuze comes out.

(4) Dispose of the fuze in accordance with unit SOP.

M102 PRACTICE STUN HAND GRENADE

A-8. The M102 (Figure A-9) is the reloadable trainer for the M84 stun hand grenade. It offers realistic characteristics of the M84 to assist in the training of missions, such as hostage rescue or capture of criminals, terrorists, or other adversaries. Table A-3 outlines the grenade's components and characteristics.

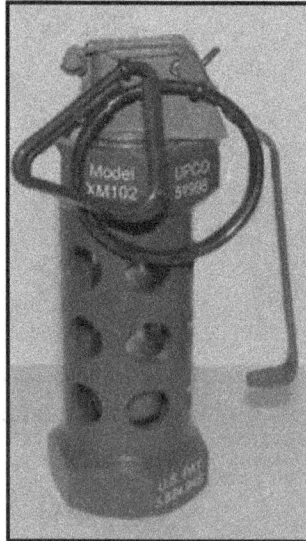

Figure A-9. M102 practice stun hand grenade.

Table A-3. Components and characteristics of M102 practice stun grenade.

COMPONENTS AND CHARACTERISTICS	DETAILS
Body	5.25 inches in length and 1.73 inches at the corner of the hexagon location; steel hexagon tube with 12 blast and flash release holes along the sides with a heavy steel, hexagon-shaped top and bottom portion
Filler	None
Fuze	M240, which is inserted into the grenade body
Safety Features	Safety pin and primary circular pull ring Safety pin and secondary triangular pull ring NOTE: The secondary triangular pull ring and safety pin will eventually be removed and a confidence clip added to the primary circular pull ring and safety pin. Safety clip (may be issued with or without a safety clip) Safety lever
Fuze Delay	1 to 2.3 seconds
Total Weight	13.3 ounces
Effects	Small puff of white smoke and a loud popping noise
Colors and Markings	Light blue with white markings; the safety lever of the fuze is light blue with black markings and a brown tip

WARNING

Fuze fragments may exit the hole in the base of the grenade body and cause injuries.

NOTE: The grenade body can be used repeatedly by replacing the fuze assembly.

M240 Detonating Fuze

A-9. Soldiers should be given instruction on installing and removing a fired M240 fuze (Figure A-10).

NOTE: Procedures for installing and removing the M240 fuze are the same as those used for the M228 fuze. For more information about these procedures, see the M228 detonating fuze section of this chapter.

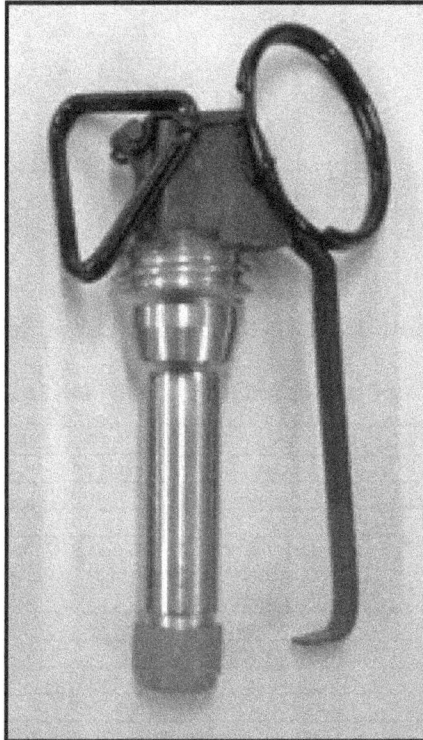

Figure A-10. M240 detonating fuze.

OFFENSIVE HAND GRENADES

A-10. Offensive hand grenades (e.g., concussion grenades) are much less lethal than fragmentation grenades on an enemy in the open, but they are very effective against an enemy within a confined space.

MK3A2 OFFENSIVE HAND GRENADE

A-11. The MK3A2 offensive hand grenade (Figure A-11) is used for concussion effects in enclosed areas, for blasting, and for demolition tasks. When used in enclosed areas, the shock waves (overpressure) produced by this grenade are greater than those produced by the fragmentation grenade. It is, therefore, more effective against enemy soldiers located in bunkers, buildings, and fortified areas. Table A-4 outlines the grenade's components and characteristics.

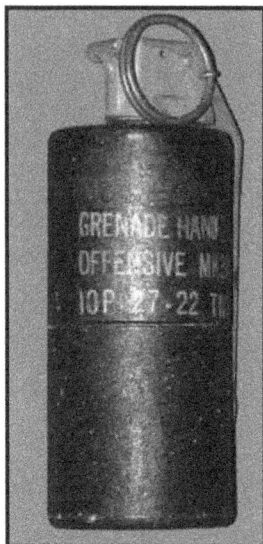

Figure A-11. MK3A2 offensive grenade.

Table A-4. Components and characteristics of MK3A2 offensive grenade.

COMPONENTS AND CHARACTERISTICS	DETAILS
Body	Fiber (similar to the packing container for the fragmentation grenade)
Filler	8 ounces of TNT
Fuze	M206A2
Safety Features	Safety clip (may be issued with or without a safety clip) Safety pin and pull ring with confidence clip Safety lever
Fuze Delay	4 to 5.5 seconds
Total Weight	15.6 ounces
Throwing Distance of Average Soldier	40 meters
Effective Casualty-Producing Radius	2 meters (in open areas)
Colors and Markings	Black with yellow markings around the middle

---

WARNING

The MK3A2 grenade has an effective casualty radius of 2 meters in open areas, but fragments and bits of fuze may be projected as far as 200 meters from the detonation point.

NONLETHAL HAND GRENADES

A-12. Nonlethal grenades are used for diversionary purposes or when lethal force is not desired. Nonlethal munitions are designed to incapacitate personnel while minimizing fatalities, permanent injury to personnel, and collateral damage to property and the environment.

M84 STUN GRENADE

A-13. The M84 stun grenade (Figure A-12) is a nonlethal grenade that is used for diversionary or distraction purposes when lethal force is not desired. Table A-5 outlines the grenade's components and characteristics.

Figure A-12. M84 stun grenade.

Table A-5. Components and characteristics of M84 stun grenade.

COMPONENTS AND CHARACTERISTICS	DETAILS
Body	5.25 inches in length and 1.73 inches at the corner of the hexagon location; steel hexagon tube with 12 blast and flash release holes along the sides with a heavy steel, hexagon-shaped top and bottom portion
Fuze	M201A1 MOD 2 (designed to be non-fragmenting)
Safety Features	Safety pin and primary circular pull ring Safety pin and secondary triangular pull ring NOTE: The secondary triangular pull ring and safety pin will eventually be removed and a confidence clip added to the primary circular pull ring and safety pin. Safety clip (may be issued with or without a safety clip) Safety lever
Fuze Delay	1 to 2.3 seconds
Total Weight	13.3 ounces
Effects	Upon detonation, the M84 generates an intensive heat, a flash of over one million candlepower, and a bang that is 170 to 180 decibels at 5 feet. The flash may damage eyesight and night vision.
Effective Casualty-Producing Radius	The grenade can cause disorientation, confusion, ear injuries, and temporary loss of hearing within 9 meters.
Colors and Markings	Olive drab with white markings, a pastel green band around the middle of the body, and a brown band on the tip end of the safety lever

CAUTION

Use stun grenades as field-expedient early warning devices only when in a combat environment.

CHEMICAL GRENADES

A-14. Chemical grenades are used for incendiary purposes or riot control.

AN-M14TH3 INCENDIARY HAND GRENADES

A-15. The AN-M14 TH3 incendiary hand grenade (Figure A-13) is used to destroy equipment or start fires. It can also damage, immobilize, or destroy vehicles, weapons systems, shelters, or munitions. Table A-6 outlines the grenade's components and characteristics.

Figure A-13. AN-M14 TH3 incendiary hand grenade.

Table A-6. Components and characteristics of AN-M14 TH3 incendiary hand grenade.

COMPONENTS AND CHARACTERISTICS	DETAILS
Body	Sheet metal
Filler	26.5 ounces of thermite (TH3) mixture
Fuze	M201A1
Safety Features	Safety clip (may be issued with or without a safety clip) Safety pin and pull ring with confidence clip Safety lever
Fuze Delay	0.7 to 2.0 seconds
Total Weight	32 ounces
Throwing Distance of Average Soldier	25 meters
Effects	A portion of thermite mixture is converted to molten iron, which burns at 4,330 degrees Fahrenheit. The mixture fuzes together the metallic parts of any object that it contacts. The thermite filler can burn through a 1/2-inch homogenous steel plate. It produces its own oxygen and burns under water.
Colors and Markings	Gray with purple markings, has a single purple band (current grenades) Under the standard color-coding system, incendiary grenades are light red with black markings.

WARNING

Avoid looking directly at the incendiary hand grenade as it burns. The intensity of the light is hazardous to the retina and can cause permanent eye damage.

RIOT-CONTROL HAND GRENADES

A-16. Riot-control hand grenades include the ABC-M7A2, ABC-M7A3, and the M47 CS.

ABC-M7A2 and ABC-M7A3

A-17. These riot-control hand grenades (Figure A-14) contain only CS fillers. The grenades differ only in the amount and form of the CS they contain. Table A-7 outlines the grenades' components and characteristics.

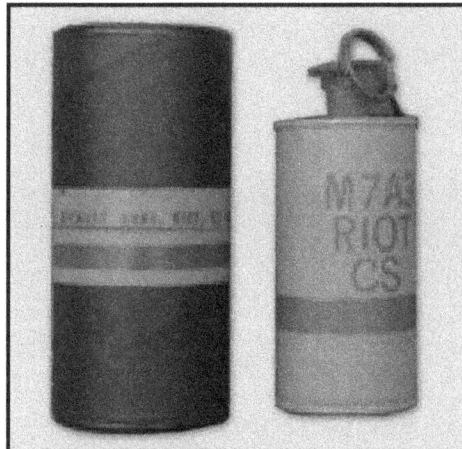

Figure A-14. ABC-M7A2 and ABC-M7A3 riot-control hand grenades.

Table A-7. Components and characteristics of ABC-M7A2 and ABC-M7A3
riot-control hand grenades.

COMPONENTS AND CHARACTERISTICS	DETAILS
Body	Sheet metal with four emission holes at the top and one at the bottom
Filler	ABC-M7A2: 5.5 ounces of burning mixture and 3.5 ounces of CS in gelatin capsules. ABC-M7A3: 7.5 ounces of burning mixture and 4.5 ounces of CS pellets
Fuze	M201A1 MOD2
Safety Features	Safety clip (may be issued with or without a safety clip) Safety pin and pull ring with confidence clip Safety lever
Fuze Delay	0.7 to 2.0 seconds
Total Weight	15.5 ounces
Throwing Distance of Average Soldier	40 meters
Effects	Both grenades produce a cloud of irritant agent for 15 to 35 seconds.
Colors and Markings	Gray bodies with red bands and markings

CAUTION

Riot-control grenades throw sparks up to 1 meter from emission, which can ignite vegetation and other flammable materials.

WARNING

Do not use a riot-control grenade in an enclosed area. If you must remain in the area, always wear a protective mask.

This page intentionally left blank.

Characteristics of Pyrotechnic Signals and Simulators

Pyrotechnic signals can be used for communication, obscuration, warning, and for simulating enemy fires. This appendix outlines the characteristics of these signals and simulators.

NOTE: A confidence clip (Figure B-1) has been introduced to pyrotechnics equipped with a safety pin and pull ring.

Figure B-1. Confidence clip.

SECTION I. COMMUNICATION SIGNALS

There are two classifications of pyrotechnic communication signals: handheld signals and ground smoke signals. Both type of signals come in varied color patterns. Soldiers can use these patterns to coordinate troop movements and, in the case of an emergency, designate pick-up points.

HANDHELD SIGNALS

B-1.   Handheld signals include:
- Star clusters.
- Star parachutes.
- Smoke parachutes.

B-2.   Star clusters, star parachutes, and smoke parachutes are issued in an expendable launcher that consists of a launching tube and firing cap (Figure B-2).

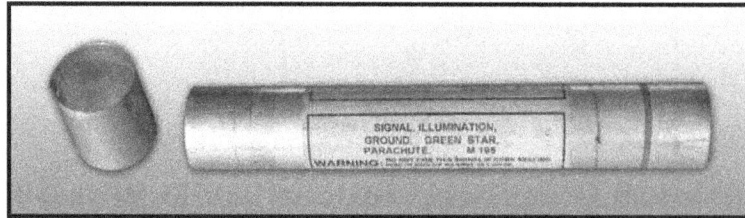

Figure B-2. Handheld pyrotechnic signal.

IDENTIFICATION

B-3.   The label and muzzle cap of a handheld ground signal identifies its color and type. The star clusters and parachutes also have two raised letters on the muzzle cap allowing the color and type to be identified at night by feel. Table B-1 provides information about handheld signal identification.

NOTE: Identification of handheld ground signals should be practiced in the dark or blindfolded.

Table B-1. Handheld signal identification.

TYPE	LETTERS ON MUZZLE CAP	COLOR OF MUZZLE CAP
Green star cluster	GS	Green
Red star cluster	RS	Red
White star cluster	WS	White
Green star parachute	GP	Green
Red star parachute	RP	Red
White star parachute	WP	White
Smoke cluster	None	Plain

STAR CLUSTERS

B-4.   Star clusters are used for signaling and illuminating. These signals produce a cluster of five free-falling pyrotechnic stars. When fired, the star cluster will rise to an approximate height of 200 to 215 meters, and burn about 6 to 10 seconds.

B-5.   Types include:
- M125 and M125A1 (green star).
- M158 (red star).
- M159 (white star).

NOTE: The white star cluster provides the most effective illumination.

STAR PARACHUTES

B-6.   Star parachutes are also used for signaling and illuminating. These signals produce a single illuminant star suspended from a parachute.

NOTE: These signals are fired in the same manner as the star clusters.

B-7.   Types include:
- M126A1 (red star).
- M127A1 (white star).
- M195 (green star).

B-8.   The M126 and M127-series star parachutes rise to a height of 200 to 215 meters. The M126 burns for 50 seconds, and the M127 burns for 25 seconds. The average rate of descent for both is 2.1 meters per second. The signals can be seen for 50 to 58 kilometers at night.

SMOKE PARACHUTES

B-9.   Smoke parachutes are for signaling only. These signals produce a single, perforated, colored smoke canister suspended from a parachute.

B-10. Types include:
- M128A1 (green smoke).
- M129A1 (red smoke).
- M194 (yellow smoke).

B-11. Smoke parachutes rise to an approximate height of 200 to 215 meters. The signals emit smoke for 12 seconds during the day, forming a smoke cloud that persists for about 60 seconds. Their rate of descent is 4 meters per second. At night, the M126A1 emits smoke for 50 seconds, the M129A1 for 25 seconds, and the M194 for 50 seconds.

GROUND SMOKE SIGNALS

B-12. These signals are self-contained units used by ground Soldiers to signal aircraft or to convey information through a prearranged signal. There are various colors of ground smoke signals.

M18 COLORED SMOKE HAND GRENADE

B-13. The M18 colored smoke hand grenade (Figure B-3) is used as a means of communication. Table B-2 outlines its components and characteristics.

Figure B-3. M18 colored smoke hand grenades.

Table B-2. Components and characteristics of M18 colored smoke hand grenade.

COMPONENTS AND CHARACTERISTICS	DETAILS
Body	Sheet steel cylinder with four emission holes at the top and one at the bottom, which allow smoke to escape when the grenade is ignited NOTE: Recently manufactured grenades do not have bottom holes.
Filler	11.5 ounces of colored smoke mixture (red, yellow, green, or violet)
Fuze	M201A1
Safety Features	Safety clip (may be issued with or without a safety clip) Safety pin and pull ring with confidence clip Safety lever
Fuze Delay	0.7 to 2.0 seconds
Total Weight	19 ounces
Throwing Distance of Average Soldier	35 meters
Effects	The grenade burns for 50 to 90 seconds with an average burn time of 60 seconds.
Colors and Markings	Light green body with black markings NOTE: The top of the grenade indicates the smoke color.

WARNING

Burning-type grenades burn oxygen. Standard protective masks filter particles but will not supply oxygen. Therefore, burning grenades must not be used in enclosed or confined spaces.

M83 TA White Smoke Hand Grenade

B-14. The M83 TA white smoke hand grenade (Figure B-4) is used for screening the activities of small units and for ground-to-air signaling. Table B-3 outlines its components and characteristics.

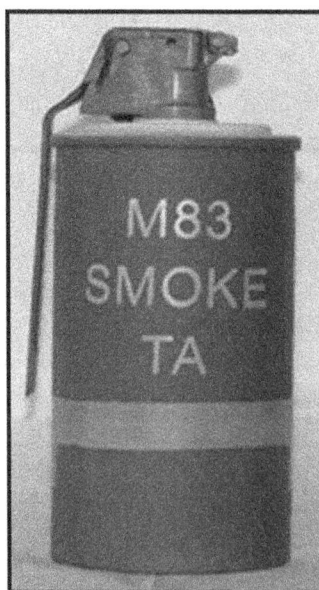

Figure B-4. M83 TA white smoke hand grenade.

Table B-3. Components and characteristics of M83 TA white smoke hand grenade.

COMPONENTS AND CHARACTERISTICS	DETAILS
Body	A cylinder of thin sheet metal, 2.5 inches in diameter and 5.7 inches long
Filler	11 ounces of terephthalic acid (TA)
Fuze	M201A1
Safety Features	Safety clip (may be issued with or without a safety clip) Safety pin and pull ring with confidence clip Safety lever
Fuze Delay	0.7 to 2.0 seconds
Total Weight	16 ounces
Effects	The M83 TA produces a cloud of white smoke for 25 to 70 seconds.
Colors and Markings	Forest green body with light green markings, a light blue band, and a white top

WARNING

Burning-type grenades burn oxygen. Standard protective masks filter particles but will not supply oxygen. Therefore, burning grenades must not be used in enclosed or confined spaces.

AN-M8 HC WHITE SMOKE HAND GRENADE

B-15. The AN-M8 HC white smoke hand grenade (Figure B-5) produces dense clouds of white smoke for signaling and screening. Table B-4 outlines its components and characteristics.

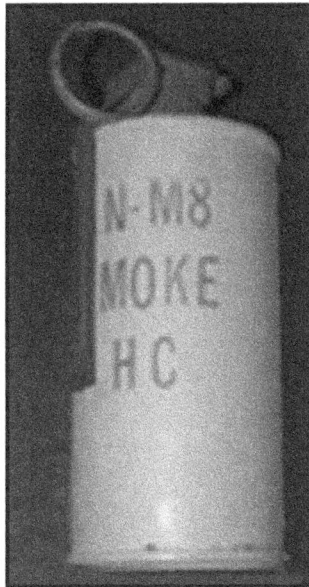

Figure B-5. AN-M8 HC white smoke hand grenade.

Table B-4. Components and characteristics of AN-M8 HC white smoke hand grenade.

COMPONENTS AND CHARACTERISTICS	DETAILS
Body	Sheet steel cylinder
Filler	19 ounces of Type C, HC smoke mixture
Fuze	M201A1
Safety Features	Safety clip (may be issued with or without a safety clip) Safety pin and pull ring with confidence clip Safety lever
Fuze Delay	0.7 to 2.0 seconds
Total Weight	24 ounces
Throwing Distance of Average Soldier	30 meters
Effects	The grenade emits a dense cloud of white smoke for 105 to 150 seconds.
Colors and Markings	Light green body with black markings and a white top

WARNING

The AN-M8 HC hand grenade produces harmful hydrochloric fumes that irritate the eyes, throat, and lungs. It should not be used in enclosed or confined spaces unless Soldiers are wearing protective masks.

Damaged AN-M8 HC grenades that expose the filler are hazardous. Exposure of the filler to moisture and air could result in a chemical reaction that will ignite the grenade.

M106 WHITE SMOKE HAND GRENADE

B-16. The M106 white smoke hand grenade (Figure B-6) provides a near instantaneous screen of dense smoke and is safe to use inside of urban structures, subterranean locations, and caves. Table B-5 outlines its components and characteristics.

NOTE: The M106 should be used in lieu of AN-M8 HC and M83 TA white smoke hand grenades when inside of confined spaces and when encountering enemy in close quarters.

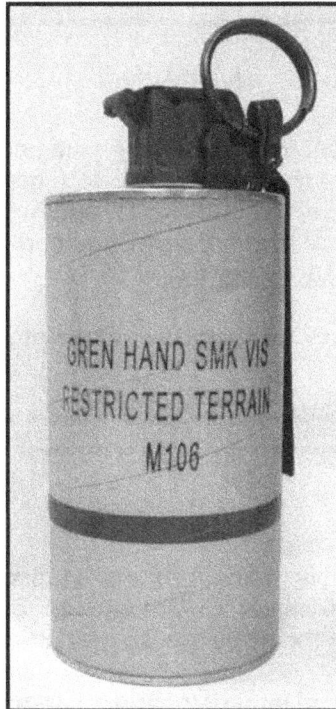

Figure B-6. M106 white smoke hand grenade.

Table B-5. Components and characteristics of M106 white smoke hand grenade.

COMPONENTS AND CHARACTERISTICS	DETAILS
Body	Fiberboard/aluminum
Filler	Non-toxic, non-combustible titanium dioxide
Fuze	M201A1 MOD 3
Safety Features	Safety clip (may be issued with or without a safety clip) Safety pin and pull ring with confidence clip Safety lever
Total Weight	20 ounces
Fuze Delay	0.7 to 2.0 seconds
Effects	The filler forms a dense, obscurant cloud within 1 to 2.3 seconds after employment. Weather effects may cause the M106 to dissipate quickly. For long-lasting smoke screens (external use only), other white smoke should be deployed in conjunction with the M106. NOTE: Inside of a building the heavy particles of the M106 may linger in the air for 2 to 4 minutes.
Colors and Markings	Light green body with black markings and a brown band, fuze is gray or olive drab with black markings

---

WARNING

The M106 smoke grenade and the M84 stun hand grenade are similar in time delay and blast effects. Both grenades can cause serious personal injury to hands, eyes, and hearing. All users must wear appropriate hearing protection and <u>exposure should be limited to two detonations per day.</u>

Do not attempt to cook off the M106 smoke grenade prior to throwing.

WARNINGS

Once the primary pin is pulled from the M106 smoke grenade, do not attempt to switch hands. Do not attempt to reinsert the primary safety pin of the M106 smoke grenade.

In the event of an accidental detonation, users must wear gloves when handling the M106 smoke grenade.

After releasing the safety lever on the M18, M83, and the AN-M8 HC smoke grenade, the firer should quickly move at least 10 meters away to avoid contact with incendiary particles and fumes emitted during burning.

CAUTION

The M106 has been tested and found to be non-toxic; however, exposure to heavy concentrations should be limited to less than 15 minutes.

SECTION II. TRIP FLARES

Surface trip flares can be used to—

- Provide early warning of infiltration of enemy troops or signaling.
- Illuminate an immediate area.
- Ignite fires.
- Identify firing ports.
- Force the enemy to withdraw.
- Destroy small, sensitive pieces of equipment (in the same manner as an incendiary grenade).

M49A1 SURFACE TRIP FLARE

B-17. Table B-6 outlines the components and characteristics of the M49A1 surface trip flare (Figure B-7).

Figure B-7. M49A1 surface trip flare.

Table B-6. Components and characteristics of M49A1 surface trip flare.

COMPONENTS AND CHARACTERISTICS	DETAILS
Body	Aluminum
Weight	0.75 pounds
Length	4.85 inches
Diameter	3.10 inches
Method of activation	Trip wire (50 feet)
Filler	Illumination composition
Primer	Percussion M42
Safety Features	The trigger is attached to the exterior of the mounting bracket. The lever is hinged to the cover and is held in position by the safety clip when unarmed. A pull on the trip wire causes either the trigger tongue or pull pin to release the lever, which in turn permits the firing pin to strike the primer. The primer sets off the intermediate charge, and the intermediate charge ignites the first-fire composition on the ignition increment of the flare.
Delay	0 seconds
Effects	The trip flare produces 35,000 candlepower illumination for 55 seconds (minimum). The area of illumination is an approximately 300-meter radius.
Colors and Markings	Olive drab body with black markings

WARNINGS

Surface trip flares can cause fires when thrown on dry tender.

The minimum safe distance from an ignited surface trip flare is 2 meters because of sparks and the popping of burning magnesium.

Never look directly at a burning surface trip flare. The intense flame can injure your eyes. At close ranges, surface trip flares may damage night vision devices and sights.

DO NOT attempt to cook off a trip flare. The fuze has a .0-second time delay.

SECTION III. SIMULATED SIGNALS

Some pyrotechnic simulators can be used to provide early warning signals and to illuminate the immediate area; however, they are primarily designed to imitate the sounds and effects of combat detonations during field training exercises.

EARLY WARNING SIMULATORS

B-18. Simulators are used in training to imitate the sounds and effects of combat detonations and the initiations of early warning devices. Booby trap simulators are activated by trip wires attached to the igniter cords, which instantaneously activate when pulled. The three types of booby trap simulators each generate a different effect upon initiation.

WARNING

Simulators are potentially dangerous if activated close to personnel or if improperly handled.

M117 Flash Explosive Booby Trap Simulator

B-19. The M117 flash booby trap simulator (Figure B-8) produces an explosion, flash, and sound. Table B-7 outlines its components and characteristics.

NOTE: The M117 simulator has a dimple in the mounting bracket for additional identification at night.

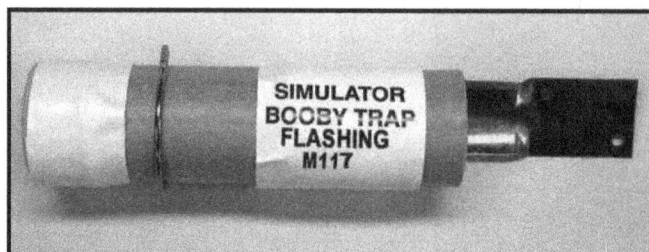

Figure B-8. M117 flash explosive booby trap simulator.

Table B-7. Components and characteristics of M117 flash explosive booby trap simulator.

COMPONENTS AND CHARACTERISTICS	DETAILS
NSN	1370-00-028-5256
Weight loaded	0.09 ounces
Length	3.9 inches
Diameter	0.98 inches
Method of activation	Tripwire
Body	Kraft paper
Color	White label with black markings
Effects	Explosion, flash, and sound
Functioning time	Instantaneous

WARNING

Do not pull the igniter cord on the M117 booby trap simulator by hand as it will immediately activate. Booby trap simulators may cause ear damage or burns if activated within 2 meters of personnel. Never open a simulator; the photoflash powder is extremely susceptible to flash ignition by even a slight amount of friction.

M118 ILLUMINATING EXPLOSIVE BOOBY TRAP SIMULATOR

B-20. The M118 illuminating booby trap simulator (Figure B-9) produces illumination. Table B-8 outlines its components and characteristics.

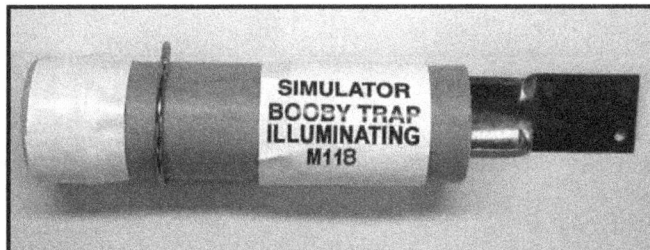

Figure B-9. M118 illuminating explosive booby trap simulator.

Table B-8. Components and characteristics of M118 illuminating explosive booby trap simulator.

COMPONENTS AND CHARACTERISTICS	DETAILS
NSN	1370-00-028-5257
Weight loaded	0.18 ounces
Length	3.9 inches
Diameter	0.98 inches
Method of activation	Trip wire
Body	Kraft paper
Color	White label with black markings
Effects	Illumination
Functioning time	28 seconds minimum flame

WARNING

Do not pull the igniter cord on the M118 booby trap simulator by hand as it will immediately activate. Booby trap simulators may cause ear damage or burns if activated within 2 meters of personnel. Never open a simulator; the photoflash powder is extremely susceptible to flash ignition by even a slight amount of friction.

M119 WHISTLING EXPLOSIVE BOOBY TRAP SIMULATOR

B-21. The M119 whistling booby trap simulator (Figure B-10) produces a whistling sound. Table B-9 outlines its components and characteristics.

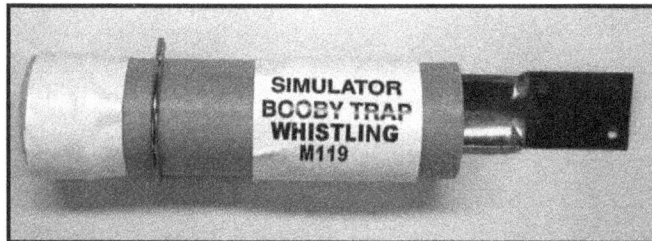

Figure B-10. M119 whistling booby trap simulator.

**Table B-9. Components and characteristics of
M119 whistling booby trap simulator.**

COMPONENTS AND CHARACTERISTICS	DETAILS
NSN	1370-00-028-5255
Weight loaded	0.12 ounces
Length	4.4 inches
Diameter	0.98 inches
Method of activation	Trip wire
Body	Kraft paper
Color	White label with black markings
Effects	Whistle sound
Functioning time	2.5 to 5 seconds

WARNING

Do not pull the igniter cord on the M119 booby trap simulator by hand as it will immediately activate. Booby trap simulators may cause ear damage or burns if activated within 2 meters of personnel. Never open a simulator; the photoflash powder is extremely susceptible to flash ignition by even a slight amount of friction.

GROUND-BURST SIMULATOR

B-22. Ground-burst simulators replicate the detonation of artillery and mortar projectiles or artillery-type rockets.

M115A2 GROUND-BURST SIMULATOR

B-23. The M115A2 ground-burst simulator (Figure B-11) is activated by pulling its M3A1 friction delay igniter cord and immediately thrown into a cleared area. After a 6- to 10-second delay, it produces a high-pitched whistle that lasts 2 to 4 seconds and then detonates with a loud report and brilliant flash. Burning pyrotechnic compound generates the whistle.

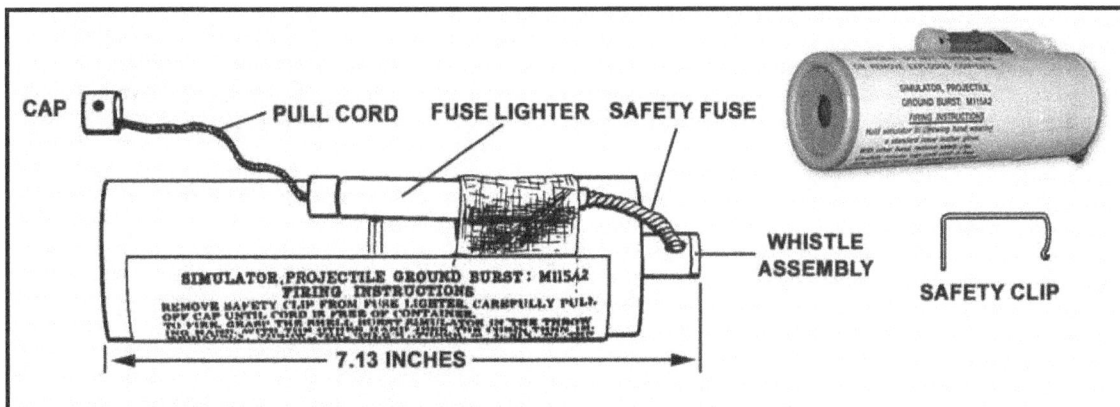

Figure B-11. M115A2 ground-burst simulator.

NOTE: Instructions for the hand grenade simulator are printed directly on the simulator.

**Table B-10. Components and characteristics of
M115A2 ground-burst simulator.**

COMPONENTS AND CHARACTERISTICS	DETAILS
NSN	1370-00-752-8126
Weight loaded	2.2 pounds
Length	7.13 inches
Diameter	2.38 inches overall
Method of activation	Hand pull cord
Body	Kraft paper
Color	White overall with white label with black markings
Effects	Whistle sound and explosive blast
Functioning time	Whistle 6 to10 seconds after ignition Burst 8 to14 seconds after ignition

WARNINGS

The M115A2 projectile ground-burst simulator must not be used near personnel due to potential hazard from fragmentation. Ensure the simulator is not thrown to any point within 35 meters of unprotected personnel.

When using the M115A2 ground-burst simulator, the thrower should turn away from the simulator after throwing and assume a protective stance.

The M115A2 must not be activated in loose gravel, sticks, or other materials that could become projectiles, nor should it be thrown into dry leaves, grass, or other flammable materials.

HAND GRENADE SIMULATOR

B-24. Hand grenade simulators are used to create battle noises and flashes during training.

M116A1 HAND GRENADE SIMULATOR

B-25. The M116A1 hand grenade simulator (Figure B-12) differs from the ground-burst simulator in that it is shorter and does not emit a high-pitched whistle before detonation. The hand grenade simulator is thrown in the same manner as a live grenade. It creates a flash and loud report 5 to 10 seconds after ignition.

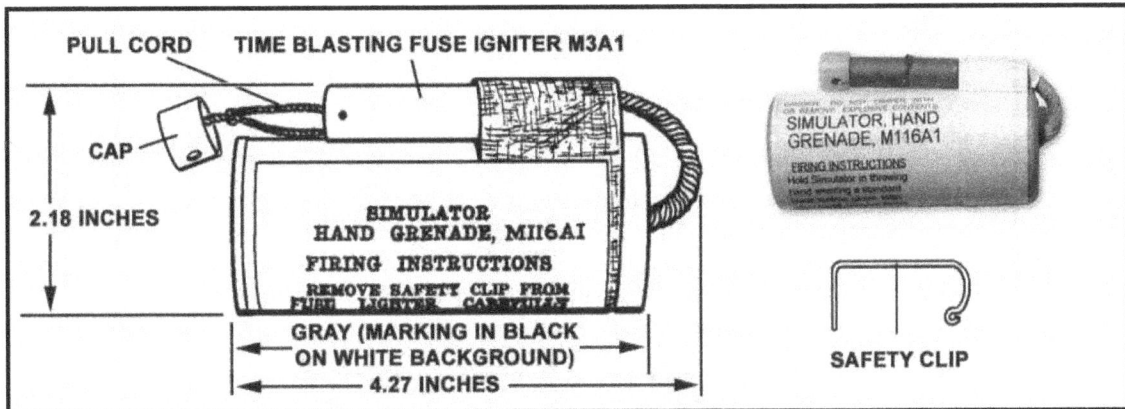

Figure B-12. M116A1 hand grenade simulator.

NOTE: Instructions for the hand grenade simulator are printed directly on the simulator.

**Table B-11. Components and characteristics of
M116A1 ground-burst simulator.**

COMPONENTS AND CHARACTERISTICS	DETAILS
NSN	1370-00-752-8124
Weight loaded	0.2 pounds
Length	4.30 inches
Diameter	2.18 inches overall
Method of activation	Hand pull cord
Body	Kraft paper
Color	White overall with white label with black markings
Effects	Explosive blast
Functioning time	Burst 5-to 10 seconds after ignition

WARNINGS

The safety radius for the M116A1 simulator is 15 meters. If this distance is not observed, hearing damage and possible fragmentation injury could result.

The M116A1 projectile ground-burst simulator must not be used near personnel due to potential hazard from fragmentation.

The M116A1 must not be activated in loose gravel, sticks, or other materials that could become projectiles, nor should it be thrown into dry leaves, grass, or other flammable materials.

SECTION IV. ILLUMINATION GROUND SIGNAL KITS

The pen gun flare supports the small-unit leader in fire control, maneuver, and initiating operations such as ambushes. These signals are also a component of air crewmen's survival vest and are used for distress signaling or to identify ground locations for aircraft. The pen gun flare comes in two kits, the M185 (red personnel signal kit) and the A/P255-5A (red personnel distress signal kit).

WARNINGS

At close-range, these signals can injure or kill if they strike a person.

When signaling an aircraft, do not aim directly at the aircraft; the signals, regardless of color, may appear to be small arms fire.

While the flares are small and usually burn out before reaching the ground, they can ignite fires.

M185 AND M186 PERSONNEL SIGNAL KITS

B-26. This pen gun flare has a threaded projector with the projectiles contained in a cloth bandoleer. The projector, the bandoleer, and seven projectiles or signals make up the signal kit (Figure B-13). All signals may be obtained and fired separately.

B-27. Four signals may be fired from a handheld projector while in a bandoleer:
- M187 red illumination ground signal.
- M188 white illumination ground signal.
- M189 green illumination ground signal.
- M190 amber illumination ground signal.

B-28. The M185 red signal kit contains seven red ground illumination signals (M187). The M186 signal kit contains three red ground illumination signals (M187), two green, and two white signals.

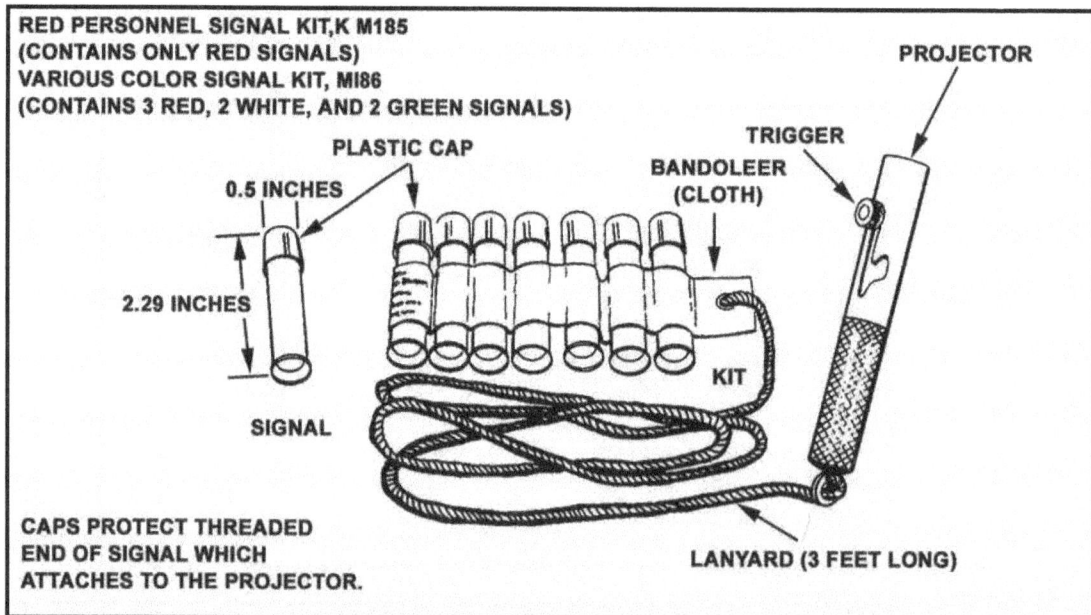

Figure B-13. M185-red/M186-various color personnel signal kit.

Table B-12. Components and characteristics of
M185-red/M186-various color personnel signal kit.

COMPONENTS AND CHARACTERISTICS	DETAILS
NSN	M185: 1370-00-921-6172 M186: 1370-00-926-9387
Weight loaded	0.39 pounds
Length	Projector: 4 inches Lanyard: 36 inches Signal: 2.29 inches
Diameter	Projector: 0.59 inches Signal: 0.5 inches
Method of activation	From projector
Body	Aluminum
Color	Black projector: anodized color coding on signals
Effects	Illuminant composition with 3200 candlepower
Functioning time	5-seconds

M260 RED PERSONNEL DISTRESS SIGNAL KIT

B-29. This pen gun flare has a force-fitted projector and a plastic bandoleer. The projector, the bandoleer, and seven signals make up this kit (Figure B-14).

B-30. This kit contains only red illumination ground signals. The signals in this kit are more powerful than those in the M185 and M186 personnel signal kits; they have a more powerful propellant allowing a higher probability of penetration through overhead foliage. The burning time for these signals is 10 seconds at 10,000 candlepower.

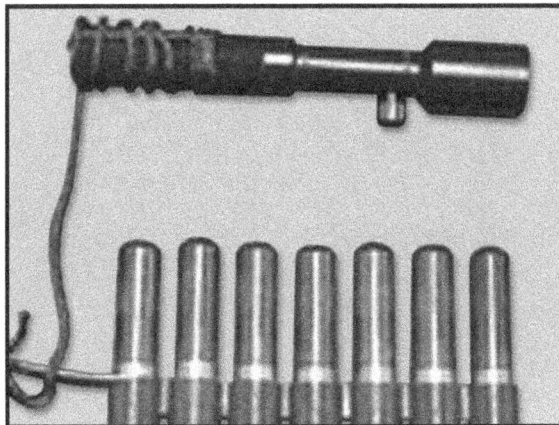

Figure B-14. M260 red personnel distress signal kit.

WARNINGS

At close-range, these signals can injure or kill if they strike a person.

When signaling an aircraft, do not aim directly at the aircraft; the signals, regardless of color, may appear to be small arms fire.

While the flares are small and usually burn out before reaching the ground, they can ignite fires.

Table B-13. Components and characteristics of M260 red personnel distress signal kit.

COMPONENTS AND CHARACTERISTICS	DETAILS
NSN	1370-00-490-7362
Weight loaded	3.2 ounces
Length	Projector: 5.5 inches Lanyard: 30 inches Signal: 2 inches
Diameter	Projector: 0.8 inches Signal: 0.5 inches
Method of activation	From projector
Body	Aluminum
Color	Black projector: anodized color coding on signals
Effects	Illuminant composition with 10,000 candlepower
Functioning time	10-seconds

Glossary

ACRONYMS AND ABBREVIATIONS

ABC	atomic, biological, and chemical
AR	army regulation
ARNG	Army National Guard
ARNGUS	Army National Guard of the United States
ASP	ammunition supply point
CALFEX	combined arms live-fire exercise
CBRN	chemical, biological, radiological, nuclear
CLP	cleaning, lubricant, and preservative
CRM	composite risk management
CS	ortho-cholorobenzalmalomonitril (irritant agent or tear gas)
DA	Department of the Army
EOD	explosive ordnance disposal
ETLBV	enhanced tactical load-bearing vest
FM	field manual
GP	general purpose
GTA	graphic training aid
HC	hexacholorethane-zinc (burning type white smoke compound)
HE	high-explosive
IET	initial entry training
LCE	load-carrying equipment
LFX	live-fire exercise
METL	mission-essential task list
METT-TC	mission, enemy, terrain (and weather), troops, time available, civil considerations
mm	millimeter
MOPP	mission-oriented protective posture
NCO	noncommissioned officer
NCOIC	noncommissioned officer in charge
NSN	national stock number
NVD	night vision device
OIC	officer in charge
PA	public address
PMCS	preventive maintenance checks and services
ROE	rules of engagement
RP	red phosphorous (casualty producing, burning-type red smoke compound)
RSO	range safety officer

SDZ	surface danger zone
SOI	signal operating/operations instructions
SOP	standing operating procedure
STX	situational training exercise
TA	teraphthalic acid (burning-type white smoke compound)
TC	training circular
TH3	thermite (burning-type incendiary compound)
TM	technical manual
TNT	trinitrotoluene
TRADOC	Training and Doctrine Command
USAIS	US Army Infantry School
USAR	US Army Reserve
WP	white phosphorous (casualty-producing, bursting-type white smoke compound)

References

DOCUMENTS NEEDED
These documents must be available to the intended users of this publication.

ARMY PUBLICATIONS
AR 385-63, *Range Safety*. 19 May 2003.
DA Pam 385-63, *Range Safety*. 4 August 2009.
FM 3-06.11, *Combined Arms Operations In Urban Terrain*. 28 February 2002.
FM 3-21.8, *The Infantry Platoon and Squad*. 28 March 2007.
FM 3-34.214, *Explosives and Demolitions*. 11 July 2007.
FM 7-0, *Training for Full Spectrum Operations*. 12 December 2008.
TC 25-8, *Training Ranges*. 5 April 2004.
TM 9-1330-200-12, *Operator's and Organizational Maintenance Manual for Grenades*. 17 September 1971.
TM 9-1370-206-10, *Operator's Manual for Pyrotechnic Signals*. 31 March 1991.
TM 9-1370-207-10, *Operator's Manual for Pyrotechnic Simulators*. 31 March 1991.

ARMY FORMS
DA Forms are available on the Army Publishing Directorate web site (www.apd.army.mil).
DA Form 2028, *Recommended Changes to Publications and Blank Forms*.
DA Form 3517-R, *Hand Grenade Qualification Scorecard*

INTERNET WEBSITES
US Army Publishing Agency, http://www.army.mil/usapa
Reimer Doctrine and Training Digital Library, http://www.adtdl.army.mil

This page is intentionally left blank.

Index

A

air operations, 5-7

C

chemical grenades, 1-1,
    A-11 to A-13 (*see also*
    incendiary hand grenade
    and riot-control hand
    grenade)

combat load factors, 1-3
    distribution, 1-3
    unit's mission, 1-3
    weight, 1-3
    weapons tradeoff, 1-3

communication signals, 1-2,
    4-1 to 4-10, B-2 to B-8 (*see
    also* handheld signals *and*
    ground smoke signals)

confidence clip, A-1 (illus),
    B-1 (illus), A-3 to A-7
    install, A-4 to A-7, A-5
    (illus), A-6 (illus), A-7
    (illus)

D

defensive operations, 5-5

destruction procedures
    early warning simulators,
        4-22
    fragmentation hand grenades,
        3-19 to 3-21
    ground-burst simulator,
        4-20 to 4-21
    handheld signals, 4-8 to 4-9
    illumination ground signal
        kits, 4-27

E

early warning simulators,
    4-16 to 4-20, B-10 to B-13
    destruction procedures, 4-20
        (*see also* destruction
        procedures)
    inspection, 4-16 to 4-17
        (*see also* inspection)
    before storing, 4-17 to 4-18,
        4-18 (illus)

simulator containers,
    4-16 to 4-17, 4-16
    (illus), 4-17 (illus)
M117 flash explosive
    booby trap simulator,
    B-10 to B-11, B-10
    (illus)
    components and
        characteristics, B-11
        (table)
M118 illuminating
    explosive booby trap
    simulator, B-11 to
    B-12, B-11 (illus)
    components and
        characteristics, B-12
        (table)
M119 whistling explosive
    booby trap simulator,
    B-12 to B-13, B-12
    (illus)
    components and
        characteristics, B-13
        (table)
maintenance, 4-19 to 4-20
    (*see also* maintenance)
storage, 4-18
    (*see also* storage)
use, 4-18 to 4-19
    (*see also* use)

employment considerations,
    1-3, 5-1 to 5-7
    (*see also* offensive
    operations, defensive
    operations, retrograde
    operations, urban
    operations, *and* air
    operations)
application, 5-1, 5-1 (tables)
considerations, 5-2
rules of engagement, 1-3,
    5-2
use under adverse
    conditions, 5-7
    limited visibility, 5-7
    MOPP4, 5-7

F

fragmentation hand grenades,
    1-1, 3-1 to 3-22, A-1 to
    A-2
    destruction procedures,
        3-20 to 3-22 (*see also*
        destruction procedures)
    inspection, 3-1 to 3-6
        (*see also* inspection)
    before storing, 3-3 to 3-5,
        3-4 (illus), (*see also*
        safety clip)
    daily checks, 3-5 to 3-6,
        3-6 (illus)
    initial inspection, 3-1 to
        3-3, 3-2 (illus), 3-3
        (illus)
M67 fragmentation hand
    grenade A-1 to A-2,
    A-2 (illus)
    components and
        characteristics, A-2
        (table)
maintenance, 3-19 to 3-20
    (*see also* maintenance)
storage, 3-6 to 3-9
    (*see also* storage)
    carrying pouch, 3-7 (illus)
    warning against taping,
        3-8 (illus)
use, 3-9 to 3-12
    (*see also* use)
    cooking off, 3-13
    gripping, 3-9 to 3-10
    left-hand grip, 3-10, 3-10
        (illus)
    right-hand grip, 3-9, 3-9
        (illus)
    preparing, 3-10 to 3-12
    pulling the safety pin,
        3-12, 3-12 (illus)
    pull ring grip, 3-11,
        3-11 (illus)
    removing the safety
        clip, 3-10 to 3-11
        (*see also* safety clip)
    throwing, 3-13 to 3-19
    standing position, 3-14
        to 3-15, 3-15 (illus)

prone-to-standing
position, 3-15 to 3-16,
3-16 (illus)
kneeling position, 3-16 to
3-17, 3-17 (illus)
prone-to-kneeling
position, 3-17 to 3-18,
3-18 (illus)
alternate prone
position, 3-18 to 3-19,
3-19 (illus)

G

ground-burst simulator, 4-20 to
4-21, B-13 to B-14, 4-21
(illus)
destruction procedures, 4-22
(*see also* destruction
procedures)
inspection, 4-21 *(see also*
inspection)
M115A2, B-13 to B-14,
B-14 (illus)
components and
characteristics, B-13 (table)
storage, 4-21 (*see also* storage)
use, 4-21 (*see also* use)

ground smoke signals, 4-9, B-3 to
B-8, 4-9 (illus)
AN-M8 HC white smoke, B-5 to
B-6, B-5 (illus)
components and
characteristics, B-6 (table)
inspection, 4-9 (*see also*
inspection)
M18 colored smoke, B-3 to B-4,
B-3 (illus)
components and
characteristics, B-4
(table)
M83 TA white smoke, B-4 to
B-5, B-4 (illus)
components and
characteristics, B-5
(table)
M106 white smoke, B-6 to B-7,
B-7 (illus)
components and
characteristics,
B-7 (table)
storage, 4-9 (*see also* storage)
use, 4-10 (*see also* use)
considerations, 4-10
employment, 4-10

gripping (*see*
fragmentation
hand grenades)
preparing (*see*
fragmentation
hand grenades)
throwing (*see*
fragmentation
hand grenades)

H

hand grenade qualification course
complete the hand grenade
qualification course,
2-33 to 2-36
course layout, 2-17 (illus)
course stations, 2-18 to 2-21,
2-33 to 2-34, 2-34
(table)
station 1, 2-18 (illus)
station 2, 2-18 (illus)
station 3, 2-18 (illus)
station 4, 2-19 (illus)
station 5, 2-20 (illus)
station 6, 2-20 (illus)
station 7, 2-21 (illus)
scoring, 2-34 to 2-36, 2-35
(illus), 2-36 (illus) (*see
also* scorecard, hand
grenade qualification)

hand grenade simulator, 4-22
to 4-23, B-14 to B-15,
4-22 (illus)
M116A1 B-14 to B-15,
B-15 (illus)
components and
characteristics, B-15
(table)

handheld signals, 4-1 to 4-8,
B-1 to B-3, 4-1 (illus),
B-2 (illus)
destruction procedures, 4-8
to 4-9
(*see also* destruction
procedures)
identification, B-2,
B-2 (table)
inspection, 4-1 to 4-5
(*see also* inspection)
signal containers, 4-2 to
4-5, 4-2 (illus), 4-3
(illus), 4-4 (illus),
4-5 (illus)

maintenance, 4-8
(*see also* maintenance)
star clusters, B-2
star parachutes, B-2
smoke parachutes, B-3
storage, 4-5 to 4-6
(*see also* storage)
use, 4-6 to 4-8 (*see also* use)
determining type and
color, 4-6
launching handheld
signals, 4-6-4-8,
4-7 (illus), 4-8 (illus)
misfire, 4-8

I

illumination ground signal
kits, 1-2, 4-23 to 4-24,
B-15 to B-18, 4-23 (illus)
destruction procedures, 4-24
(*see also* destruction
procedures)
inspection, 4-23
(*see also* inspection)
M185 and M186 personnel
signal kits, B-16 to
B-17, B-16 (illus)
components and
characteristics,
B-17 (table)
M260 red personnel
distress signal kit,
B-17 to B-18,
B-17 (illus)
components and
characteristics,
B-18 (table)
maintenance, 4-24
(*see also* maintenance)
storage, 4-24
(*see also* storage)
use, 4-24 (*see also* use)

incendiary hand grenade,
A-11 to A-12
AN-M14TH3, A-11 to
A-12, A-12 (illus)
components and
characteristics,
A-12 (table)
gripping (*see*
fragmentation hand
grenades)
preparing (*see*
fragmentation hand

grenades)
throwing (*see*
fragmentation hand
grenades)

inspection
early warning simulators,
4-15 to 4-18
fragmentation hand
grenades, 3-1 to 3-6
ground-burst simulator,
4-21
ground smoke signals, 4-9
handheld signals, 4-1 to 4-5
illumination ground signal
kits, 4-23

M

M69 practice hand grenade,
A-2 to A-7, A-3 (illus)
components and characteristics,
A-3 (table)
gripping (*see* fragmentation
hand grenades)
M228 detonating fuze, A-3,
A-4 (illus)
install, A-4 to A-7,
A-5 (illus), A-6 (illus),
A-7 (illus) (*see also*
confidence clip)
remove, A-7
preparing (*see* fragmentation
hand grenades)
throwing (*see* fragmentation
hand grenades)

M102 practice stun hand grenade,
A-8 to A-9, A-8 (illus)
components and characteristics,
A-8 (table)
gripping (*see* fragmentation
hand grenades)
M240 detonating fuze,
A-9, A-9 (illus)
preparing (*see* fragmentation
hand grenades)
throwing (*see* fragmentation
hand grenades)

maintenance
early warning simulators, 4-19 to
4-20
fragmentation hand
grenades, 3-19
handheld signals, 4-8

illumination ground signal
kits, 4-24
trip flares, 4-15

N

nonlethal hand grenade, 1-2,
A-10 to A-11
M84 stun grenade, A-10 to
A-11, A-11 (illus)
components and
characteristics, A-11
(table)
gripping (*see*
fragmentation
hand grenades)
preparing (*see*
fragmentation
hand grenades)
throwing (*see*
fragmentation
hand grenades)

O

offensive hand grenade, 1-2,
A-9 to A-10
MK3A2 offensive hand
grenade, A-9 to A-10,
A-10 (illus)
components and
characteristics,
A-10 (table)
gripping (*see*
fragmentation hand
grenades)
preparing (*see*
fragmentation hand
grenades)
throwing (*see*
fragmentation hand
grenades)

offensive operations, 5-2 to 5-5
clearing a room, 5-3 to 5-4
clearing bunkers, 5-3
confined spaces, 5-2 to 5-3
entrenched position, 5-4 to
5-5
trenches, 5-3

P

practice and training
grenades, 1-2, A-2 to A-7
(*see also* M69 practice hand
grenade *and* M102 practice
stun hand grenade)

pyrotechnic signals and
simulators, 1-2, 4-1 to
4-24, B-1 to B-18
communication signals,
4-1 to 4-10, B-2 to B-8
(*see also* communication
signals)
illumination ground signal
kits, 4-23 to 4-24, B-15 to
B-18 (*see also*
illumination ground
signal kits)
simulated signals, 4-15 to
4-22, B-12 to B-15 (*see
also* simulated signals)
trip flares, 4-11 to 4-15, B-8
to B-10 (*see also* trip
flares)

R

range personnel, 2-21 to 2-25
ammunition personnel, 2-25
guards, 2-25
medical personnel, 2-25
officer in charge and
noncommissioned officer
in charge, 2-22,
2-22 (table)
pit safety
noncommissioned
officers, 2-22 to 2-25
drop procedures,
2-23 to 2-25
soldier drops the hand
grenade, 2-24 to 2-25
soldier fails to take
commands, 2-25
soldier freezes after
arming the hand
grenade, 2-24
soldier milks the hand
grenade, 2-23
soldier remains
standing after he
throws the hand
grenade downrange,
2-24
range safety officer, 2-23,
2-22 (table)
tower operator, 2-25
truck driver, 2-25

ranges, 2-10 to 2-21
distance and accuracy
ranges, 2-11, 2-11 (illus)

hand grenade qualification course, 2-17 to 2-21 (*see also* hand grenade qualification course)
live-bay throwing pits, 2-12 to 2-16, 2-13 (illus), 2-14 (illus), 2-15 (illus), 2-16 (illus)
mock-bay throwing pits, 2-12, 2-12 (illus)

retrograde operations, 5-5 to 5-6
break contact, 5-6
communications, 5-6
create obstacles, 5-6
mark locations, 5-6

riot-control hand grenade, A-12 to A-13
ABC-M7A2 and ABC-M7A3-14 A-14, A-13 (illus)
components and characteristics, A-13 (table)
gripping (*see* fragmentation hand grenades)
preparing (*see* fragmentation hand grenades)
throwing (*see* fragmentation hand grenades)

S
safety clip, 3-4 to 3-5
installation, 3-4 to 3-5
removing the safety clip, 3-10 to 3-13
left hand grip, 3-11 (illus)
right hand grip, 3-10 (illus)

scorecard, hand grenade qualification, 2-34 to 2-36
DA Form 3517-R, 2-35 (illus), 2-36 (illus)

simulated signals, 1-2, 4-15 to 4-20, B-10 to B-15 (*see also* early warning simulators, ground-burst simulator *and* hand grenade simulator)

storage
early warning simulators, 4-18 to 4-19
fragmentation hand grenades, 3-6 to 3-9
ground-burst simulator, 4-21
ground smoke signals, 4-9
handheld signals, 4-5 to 4-6
illumination ground signal kits, 4-24
trip flares, 4-13

T
training, 2-1 to 2-43
training conduct, 2-28 to 2-43 (*see also* training, conduct)
training preparation, 2-8 to 2-28 (*see also* training, preparation)
training program, 2-3 to 2-8 (*see also* training, program)
training strategy, 2-1 to 2-3 (*see also* training, strategy)

training, assessment, 2-3 to 2-5
commander's evaluation guide, 2-4 to 2-5
commander's priorities and intent, 2-4 to 2-5
soldier assessment, 2-5
trainer assessment, 2-5
direct observation of training, 2-4
spot checks, 2-4
review of past training, 2-4

training, conduct, 2-27 to 2-43
complete the training mission, 2-43
conduct the training, 2-27 to 2-43
collective training, 2-42 to 2-43
squad situational training exercise, 2-42 to 2-43, 2-43 (illus)
initial training, 2-28 to 2-36 (*see also* training, initial)
occupy, inspect, and set up

range, 2-27 to 2-28
preparation for training, 2-28 (*see also* training, preparation)
sustainment training, 2-36 to 2-42 (*see also* training, sustainment)

training, initial, 2-1 to 2-2, 2-28 to 2-36
complete the hand grenade qualification course, 2-33 to 2-36 (*see also* hand grenade qualification course)
participate in distance and accuracy training, 2-29, 2-29 (table)
participate in initial hand grenade training, 2-28
participate in live-bay training, 2-31 to 2-33
participate in mock-bay training, 2-29 to 2-31

training, preparation, 2-8 to 2-27
conduct an environmental risk assessment, 2-10
assess hazards, 2-10
identify hazards, 2-10
conduct a training risk assessment, 2-8 to 2-10
assess hazards to determine risk, 2-9
develop controls and make risk decisions, 2-9
identify hazards, 2-10
implement controls, 2-10
supervise and evaluate, 2-10
make range coordinations 2-10 to 2-27
duds 2-26 to 2-27, 2-27 (table)
equipment 2-21
personnel 2-21 to 2-26, 2-22 (table) (*see also* range personnel)
ranges 2-10 to 2-21 (*see also* ranges)
surface danger zone 2-25, 2-25 (illus)

training, program, 2-3 to 2-8
mission-essential tasks, 2-3
trainers, 2-5 to 2-7
cadre/trainer, 2-5 to 2-6
duties, 2-6
selection, 2-6
training the trainer, 2-6 to 2-7
training assessment, 2-3 to
2-5 (*see also* training,
assessment)
trainer certification
program, 2-7 to 2-8
certification program
outline, 2-7 to 2-8
training base, 2-7

training, strategy 2-1 to 2-3
initial and sustainment
training 2-1 to 2-3
initial training 2-1 to 2-2, 2-2
(illus) (*see also* training,
initial)
sustainment training 2-2 to 2-3
(*see also* training,
sustainment)
objectives 2-1

training, sustainment, 2-37 to
2-42
complete the building

complex course, 2-39,
2-39 (table)
complete the bunker
complex course, 2-39,
2-39 (table)
complete the distance and
accuracy course, 2-38,
2-38 (table)
complete the trench
complex course, 2-39 to
2-40, 2-40 (table)
employ a pyrotechnic
signal, 2-38, 2-38 (table)
identify hand grenades and
pyrotechnic signals, 2-37,
2-37 (table)
inspect and maintain hand
grenades and pyrotechnic
signals, 2-37, 2-37 (table)
participate in live-bay
training, 2-42, 2-42
(table)
participate in mock-bay
training, 2-41, 2-41
(table)

trip flares, 1-2, 4-11 to 4-15,
B-8 to B-10, 4-11 (illus)
M49A1 surface trip flare,
B-8 to B-10, B-9 (illus)
components and

characteristics, B-9
(table)
maintenance, 4-15 (*see also*
maintenance)
mounting bracket, 4-12,
4-12 (illus)
removal, 4-15
storage, 4-13
(*see also* storage)
use, 4-13 to 4-15
(*see also* use)
hand-thrown, 4-14 to 4-15
mounted, 4-13 to 4-14

U
urban operations, 5-6 to 5-7,
5-6 (table)

use
early warning simulators,
4-18 to 4-19
fragmentation hand
grenades, 3-9 to 3-12
ground-burst simulator,
4-21
ground smoke signals,
4-10
handheld signals, 4-6 to 4-8
illumination ground signal
kits, 4-24
trip flares, 4-13 to 4-15

This page is intentionally left blank.

HAND GRENADE QUALIFICATION SCORECARD

For use of this form, see FM 3-23.30; the proponent agency is TRADOC

NOTE: In addition to the requirements on the scorecard, the Soldier must throw two live fragmentation grenades to qualify.

A. NAME *(Last, First, Middle Initial)*	B. UNIT		C. DATE *(YYYYMMDD)*
D. EVALUATOR'S NAME		E. DATE LIVE GRENADES WERE THROWN *(YYYYMMDD)*	

F. STATION	G. TYPE TARGET	H. GO	I. NO-GO	J. EVALUATOR'S INITIALS
1	Engage Enemy from Fighting Position at 35 Meters *(Standing)*	☐	☐	
2	Engage Bunker *(Prone)*	☐	☐	
3	Engage Enemy Mortar Position at 25 Meters *(Kneeling)*	☐	☐	
4	Engage Enemy Behind Cover at 20 Meters *(Prone)*	☐	☐	
5	Engage Trench at 25 Meters *(Standing)*	☐	☐	
6	Engage Wheeled Vehicle at 25 Meters *(Kneeling)*	☐	☐	
7	Identify Hand Grenades and Pyrotechnic Signals	☐	☐	

K. QUALIFICATION STANDARD		CHECK ONE
PASSED 7	EXPERT	☐
PASSED 6	FIRST CLASS	☐
PASSED 5	SECOND CLASS	☐
PASSED 4 OR LESS	UNQUALIFIED	☐

L. DATE INITIALED *(YYYYMMDD)*	M. SCORER'S INITIALS
N. DATE INITIALED *(YYYYMMDD)*	O. SCORER'S INITIALS

DA FORM 3517-R, NOV 2009   PREVIOUS EDITION IS OBSOLETE.   APD PE v1.00

PERFORMANCE MEASURES	GO	NO-GO	PERFORMANCE MEASURES	GO	NO-GO
STATION 1. Engage Enemy From Fighting Position at 35 Meters *(Standing)*			**STATION 5. Engage Trench at 25 Meters** *(Standing)*		
A. Detonated at least one grenade within 5 meters of the center of target.	☐	☐	A. Detonated at least one grenade inside trench.	☐	☐
B. Kept exposure time under 3 seconds.	☐	☐	B. Kept exposure time under 3 seconds.	☐	☐
C. Returned to covered position after each throw.	☐	☐	C. Returned to covered position after each throw.	☐	☐
D. Used proper grip.	☐	☐	D. Used proper grip.	☐	☐
E. Used proper throwing techniques.	☐	☐	E. Used proper throwing techniques.	☐	☐
F. Completed performance measures 1A through 1E within 15 seconds.	☐	☐	F. Completed performance measures 5A through 5E within 15 seconds.	☐	☐
STATION 2. Engage Bunker *(Prone)*			**STATION 6. Engage Wheeled Vehicle at 25 Meters** *(Kneeling)*		
A. Approached from blind side.	☐	☐	A. Detonated within 1 meter of vehicle or within 5 meters of dismounting troops.	☐	☐
B. Checked for bunker opening.	☐	☐	B. Kept exposure time under 3 seconds.	☐	☐
C. Detonated grenade in bunker.	☐	☐	C. Returned to covered position after each throw.	☐	☐
D. Rolled away from bunker.	☐	☐	D. Used proper grip.	☐	☐
E. Used proper grip.	☐	☐	E. Used proper throwing techniques.	☐	☐
F. Used cook-off technique.	☐	☐	F. Completed performance measures 6A through 6E within 15 seconds.	☐	☐
G. Completed performance measures 2A through 2F within 15 seconds.	☐	☐	**STATION 7. Identify Hand Grenades and Pyrotechnic Signals**		
STATION 3. Engage Mortar Position at 25 Meters *(Kneeling)*			A. Selected fragmentation grenade to engage enemy soldiers.	☐	☐
A. Detonated at least one grenade inside mortar position.	☐	☐	B. Identified M83 grenade as "White Smoke" or "HC Smoke."	☐	☐
B. Kept exposure time under 3 seconds.	☐	☐	C. Identified M18 grenades as "Colored Smoke" or "Purple *(and so forth)* Smoke." *(If specific color is stated, it must be the same as color on the training aid grenade used.)*	☐	☐
C. Returned to covered position after each throw.	☐	☐	D. Identified M7A2/A3 grenade as CS or riot control.	☐	☐
D. Used proper grip.	☐	☐	E. Identified M14 grenades as incendiary.	☐	☐
E. Used proper throwing techniques.	☐	☐	**NOTES:**		
F. Completed performance measures 3A through 3E within 15 seconds.	☐	☐	1. FOR PERFORMANCE MEASURES 7A THROUGH 7E, IF THE EXAMINEE CANNOT CORRECTLY STATE THE NAME OF THE GRENADE, BUT CAN CORRECTLY IDENTIFY ITS USE, THEN THE EXAMINEE WILL BE SCORED A "GO."		
STATION 4. Engage Enemy Behind Cover at 20 Meters *(Prone)*			2. EACH PERFORMANCE MEASURE AT EACH SECTION IS GRADED ON A PASS/FAIL STANDARD. A SOLDIER MUST PASS ALL OF THE STANDARDS TO RECEIVE A "GO" ON THAT STATION.		
A. Detonated at least one grenade within 5 meters of the center of target.	☐	☐			
B. Kept exposure time under 3 seconds.	☐	☐			
C. Returned to covered position after each throw.	☐	☐			
D. Used proper grip.	☐	☐			
E. Used proper throwing techniques.	☐	☐			
F. Completed performance measures 4A through 4E within 15 seconds.	☐	☐			

By order of the Secretary of the Army:

GEORGE W CASEY, JR.
General, United States Army
Chief of Staff

Official:

JOYCE E. MORROW
Administrative Assistant to the
Secretary of the Army
0923702

DISTRIBUTION:

Active Army, Army National Guard, and U.S. Army Reserve: To be distributed in accordance with the initial distribution number (IDN) 110196, requirements for FM 3-23.30.

www.ingramcontent.com/pod-product-compliance
Lightning Source LLC
Chambersburg PA
CBHW051218200326
41519CB00025B/7157